Exercise and Sports Pulmonology

Annalisa Cogo · Matteo Bonini
Paolo Onorati
Editors

Exercise and Sports Pulmonology

Pathophysiological Adaptations and Rehabilitation

Springer

Editors
Annalisa Cogo
Centre for Exercise and Sport Science
University of Ferrara
Ferrara
Italy

Matteo Bonini
National Heart and Lung Institute
Imperial College London
London
UK

Paolo Onorati
Internal medicine Department,
Ospedale Civile di Alghero
ASSL Sassari, ATS Sardegna
Alghero
Italy

ISBN 978-3-030-05257-7 ISBN 978-3-030-05258-4 (eBook)
https://doi.org/10.1007/978-3-030-05258-4

Library of Congress Control Number: 2019930437

This Springer imprint is published by the registered company Springer Nature Switzerland AG
The registered company address is: Gewerbestrasse 11, 6330 Cham, Switzerland

Foreword

Exercise is arguably the cardinal stressor to the lungs, whether in the elite athlete or the highly compromised patient with chronic respiratory disease. This is the ambitious theme of *Exercise and Sports Pulmonology: Pathophysiological Adaptations and Rehabilitation*, for which one should perhaps be mindful of Joseph Barcroft's (1872–1947) aphorism that "The condition of exercise is not a mere variant of the condition of rest, it is the essence of the machine" (*Features in the Architecture of Physiological Function*. Cambridge University Press, p 286, 1934). And while the "*machine*" itself is common to both athlete and patient, the systemic manifestations of its "*essence*" that ultimately lead to exercise intolerance naturally differ widely; as do the means at hand to extend their tolerable system limits for sporting accomplishment and for the maintenance of activities of daily living. The Editors—Annalisa Cogo, Matteo Bonini, and Paolo Onorati—have attracted a cadre of international investigators who insightfully address this broad continuum of human performance from a perspective of physiological and pathophysiological system integration for the athlete, the patient, the athlete as patient, and the patient as athlete.

Susan A. Ward
University of Leeds,
Leeds, UK

Preface

Lungs play a pivotal role in exercise, and respiratory impairment, due to both acute and chronic lung diseases, can have a significant impact on physical performance. Furthermore, recent evidence shows that exercise is an effective and safe intervention for prevention and rehabilitation in respiratory diseases. As pulmonologists interested in "lung and exercise" and as sports physicians focused on "exercise and lung," we perceived the lack of a text specifically dedicated to this topic. We therefore thought of a text that could gather worldwide experts in a broad variety of topics concerning the relationship between lung and exercise. When the Springer's invitation arrived to plan a text on sport and lung, we accepted immediately.

We started working on a text that included the adaptations of the respiratory system during exercise, the clinical and functional assessment of subjects affected by respiratory diseases, and the eligibility and limitations to the execution of sport activity in chronic respiratory diseases with particular attention to the diagnosis, prevention, and treatment of exercise-induced bronchoconstriction also in view of the World Anti-Doping Agency rules. We included also sections concerning respiratory muscle training and rehabilitation and a section dealing with the relationships between the environment and sports. This book provides an updated overview on the link between lungs and exercise, both in healthy active subjects and in those with chronic respiratory diseases. As the first comprehensive text on this topic, we do hope it could raise reader's interest and fill a relevant unmet need.

We thank Springer for the innovative project and for the trust conferred and all the authors for having adhered enthusiastically and with dedication to our proposal.

Ferrara, Italy — Annalisa Cogo
London, UK — Matteo Bonini
Alghero, Italy — Paolo Onorati

Contents

Part I

Exercise Lung Physiology

Pulmonary Physiology and Response to Exercise

1

Carli M. Peters and A. William Sheel

Abstract

The pulmonary system is a key element within the integrated network responsible for maintaining blood gas homeostasis. In all physiologic states, including rest, sleep, and dynamic exercise, its primary function is to ensure that mixed venous blood is transformed into arterial blood with appropriate partial pressures of O_2 and CO_2. This system faces several significant challenges to its ability to maintain blood gas homeostasis during dynamic exercise. With a simultaneous increase in mixed venous carbon dioxide content and decrease in mixed venous oxygen content, ventilatory demands are significantly greater during exercise than during rest. A reduction in transit time of red blood cells through the pulmonary capillaries, resulting from an increased cardiac output, also decreases the time available for gas exchange. To maintain blood gas homeostasis despite these challenges, medullary neural networks and sensory reflex mechanisms tightly regulate alveolar ventilation. The structural capacity for producing ventilation and increasing diffusion surface area to meet the demands of dynamic exercise in the healthy respiratory system is truly remarkable.

1.1 Introduction

The human respiratory system serves multiple functions. First, this system is responsible for maintaining blood gas homeostasis by precisely matching the level of ventilation to the metabolic requirement of oxygen delivery and carbon dioxide elimination. As a given level of ventilation can be achieved in several ways, the

C. M. Peters · A. W. Sheel (✉)
School of Kinesiology, The University of British Columbia, Vancouver, BC, Canada
e-mail: bill.sheel@ubc.ca

A. Cogo et al. (eds.), *Exercise and Sports Pulmonology*,
https://doi.org/10.1007/978-3-030-05258-4_1

respiratory system must fine-tune the breathing pattern so that maintenance of blood gases is achieved at a minimum energy cost. Second, by regulating the body's CO_2 stores, the lung maintains an appropriate acid-base balance. Lastly, the lung receives the entire cardiac output, and pulmonary vascular pressure and resistance must remain low to prevent increasing the load on the right heart or damaging the thin alveolar-capillary interface. At rest and during moderate-intensity exercise, the structural and functional capacities of the respiratory system generally exceed the demands placed on the system, and the aforementioned functions are readily achieved. However, significant pulmonary limitations to exercise performance and oxygen delivery may occur during high-intensity exercise. This chapter focuses on the structure and function of the healthy human respiratory system and how the increased ventilatory requirements of exercise are met.

1.2 The Structure of the Human Respiratory System

The lungs, airways, rib cage, and respiratory muscles comprise the respiratory system. The design of the human lung is such that it permits air and blood to be in close proximity over a very thin (about 0.3 μm in some places), large (approximately 50–100 m) surface area. Its primary function is gas exchange, to allow oxygen to move from inspired air into the blood and carbon dioxide to move out. Fick's law of diffusion states that the amount of gas that moves through a sheet is inversely proportional to the thickness and proportional to the area; therefore, the structure of the blood-gas barrier is ideally suited for gas exchange. The airways are a series of branching tubes that from proximal to distal decrease in radius and length and become more numerous. Though multiple models of the human airway tree have been proposed, the most widely accepted model is that of Weibel, which numbers successive airway generations from the trachea (generation 0) down to the alveoli (generation 23) [1]. A regular dichotomy, with each bronchus regularly dividing into two daughter bronchi of approximately equal size, is assumed with this model. Three-dimensional reconstruction of computed tomography scans has demonstrated that a regular dichotomy occurs until the sixth generation of airways, but trifurcations and airways that terminate early may occur beyond this point [2]. The lungs and airways are protected within the chest cavity by the rib cage. The rib cage consists of the sternum and 12 pairs of ribs anchored posteriorly to the 12 thoracic vertebrae. The respiratory muscles modify the volume of the chest wall to produce the pressures required for inspiratory and expiratory flow generation.

An appropriate level of ventilation is maintained by the coordinated contraction of upper airway muscles and the inspiratory and expiratory muscles. Due to the negative pleural pressure developed during inspiration, muscles of the upper airways are responsible for stiffening and dilating the upper airways to ensure they don't collapse. The primary muscle of inspiration is the diaphragm. When the diaphragm contracts, the abdominal contents are pushed downward and forward and the chest cavity increases in size. Other inspiratory muscles include the external intercostals, scalene muscles, and sternocleidomastoids. Upon contraction, the external intercostals pull the ribs upward and forward, the scalene muscles elevate the first two ribs, and the sternocleidomastoid raises the sternum. The primary expiratory muscles are the abdominal muscles, including the rectus abdominis, internal and external oblique muscles, and the transverse abdominis. Intra-abdominal pressure increases upon contraction of these expiratory muscles and pushes the diaphragm upward. The internal intercostals pull the ribs downward and inward and assist with expiration.

1.3 Control of Breathing

Precise regulation of the respiratory pattern requires a control system that is capable of generating a respiratory rhythm, ensuring that the motor output to respiratory muscles is timed appropriately, and receiving sensory feedback from the periphery [3]. It is generally agreed upon that the generation of a respiratory rhythm is controlled by the central pattern generator that is made up of groups of neurons located in the pons and the medulla. Within the ventrolateral medulla, there is a group of cells known as the pre-Botzinger complex that appears essential for generating the respiratory pattern [4]. In the medulla, a group of cells referred to as the dorsal respiratory group is mainly associated with inspiration, and another group of cells, the ventral respiratory group, is associated with expiration. The respiratory rhythm is passed from these neurons to the phrenic, intercostal, and abdominal motor nerves via the spinal cord allowing the respiratory muscles to generate an appropriate pressure. To ensure proper timing of motor output to the respiratory muscles, premotor neurons are present in the medullary pattern generator that know what other neurons are doing throughout the breathing cycle [5].

1.3.1 Chemical and Mechanical Sensory Inputs

The respiratory pattern generator receives several types of feedback to fine-tune respiration. Humans have two types of chemoreceptors: one located in the medulla and bathed in brain interstitial fluid and the other located near the bifurcation of the carotid artery and exposed to arterial blood. A highly regulated ion composition is maintained in the cerebral fluid surrounding the medullary chemoreceptors due to the selective permeability of cerebral blood vessels. This selective permeability results in carbon dioxide easily entering the interstitial fluid and quickly changing the pH, whereas metabolic acids and bases in the plasma enter the brain interstitial fluid very slowly. Central chemoreceptors respond to changes in PCO_2, and though they don't respond as rapidly as peripheral receptors, after several minutes of increased PCO_2, the central chemoreceptors are responsible for most of the increase in ventilation. Peripheral chemoreceptors are small organs that respond rapidly to changes in pH, PCO_2, and PO_2 in blood on the way to the brain. These small organs receive the highest blood flow per gram of tissue of any organ in the body. The carotid sinus nerve relays sensory information from the carotid bodies to stimulate the brainstem medullary respiratory neurons and shape motor nerve output to the respiratory muscles. The respiratory center also receives feedback on the mechanical state of the lung. The lung is innervated by vagal afferents that during inspiration and activation of pulmonary stretch receptors send feedback to the medulla via the vagus nerve (cranial nerve X) to inhibit inspiration. In response to low lung volumes, receptors in the lung can also provide excitatory input to the respiratory neurons. The diaphragm and abdominal muscles, similar to other skeletal muscles, also contain receptors classified as primarily mechanical (type III) and primarily metabolic (type IV) [6]. Contraction-induced mechanical and chemical stimuli activate these receptors and send feedback to the medulla via afferent fibers within the major motor nerves [7]. Feedback from these receptors plays a major role in controlling both ventilation and circulation during exercise.

1.4 Breathing Mechanics

Contraction of the inspiratory muscles reduces the pressure within the alveoli compared to atmospheric pressure and establishes the pressure gradient required to generate airflow into the lungs. Both the resistance to flow through the airways and the elastic recoil of the lung must be overcome by the inspiratory muscles during inspiration. Resistance to flow through the airways, governed by Poiseuille's law, is primarily dependent on airway radius. As lung volume is increased, tethering of the airways leads to a reduction in resistance to flow. The elastic recoil of the lung is such that without the outward recoil of the chest wall, the lung would collapse to its smallest volume. Elastic tissues such as collagen and elastin, as well as surface tension of the liquid film lining the alveoli, contribute to the lung's elastic behavior. At functional residual capacity (FRC), defined as the volume of air in the lung after a normal expiration, elastic recoil of the lung is balanced by the outward recoil of the chest wall. The elastic nature of the lung at any given volume is defined by compliance, the change in lung volume for a given pleural pressure change (Δvolume/Δpressure). The compliance of the lung is about 200 mL cm H_2O^{-1} within the range of inflation pressures generated during resting tidal breathing (between −5 and −10 cm H_2O) but is much stiffer and less compliant at high lung volumes. During dynamic exercise, the pressure, volume, and flow rate demands placed on the respiratory system are significantly increased, and precise regulation of the resistive and elastic behavior of the system is required to prevent excessive respiratory muscle work.

1.4.1 Control of Airway Caliber

The airways can be categorized as extra- or intrathoracic with the nasal, pharyngeal, and laryngeal airways considered to be the primary extrathoracic or "upper airways." For a detailed treatment of the anatomy and physiology of the upper airway, the reader is directed to Dempsey et al. [8]. The extrathoracic airways provide the majority of total respiratory system flow resistance. During strenuous exercise, there is a significant negative intrathoracic pressure, and this downstream pressure renders the extrathoracic airways particularly susceptible to narrowing during inspiration. Fortunately, the change in airflow from a nasal route (high resistance) to an oral route (low resistance) that occurs at minute ventilations of approximately 20–40 L/min serves to reduce airway resistance. Activation of the upper airway musculature during inspiration provides a “stiffening” of the extrathoracic airways which provides “traction” to abduct and dilate the upper airways [9]. The mechanisms by which activation of the upper airway muscles modulate airway resistance are unclear. However, there is some indirect evidence that activation occurs via feedforward [10] and feedback mechanisms [11].

The intrathoracic airways are susceptible to narrowing via constriction of bronchial smooth muscle. However, in humans free from respiratory disease, it is known that intrathoracic resistance is lower during and following whole-body exercise [12]. Dilation of the bronchi with exercise occurs via a reduction in cholinergic tone to the airway smooth muscle [13, 14]. There is also a significant mechanical aspect to exercise-induced bronchodilation whereby lung stretch at a high tidal volume “tethers” the airways [15].

1.4.2 Flow-Volume Relationships

During resting breathing (or eupnea), the airways expand slightly on inspiration and during quiet expiration return to their normal diameter. Expiration is typically longer than inspiration, and the average flow rates are generally smaller during expiration. When expiration is forced, airway mechanics and associated flows differ substantially from inspiration. A maximal flow-volume curve for an individual can be generated by having the subject inspire until total lung capacity (TLC), expire as hard and fast as they can to residual volume (RV), and then immediately forcefully inspire back to TLC. The resulting flow-volume curve has a characteristic shape in healthy young subjects, which can be seen in Fig. 1.1. As lung volume falls during the forced expiration, flow rate also falls considerably. This differs from the forced inspiration where flow rate remains relatively constant and results in the average flow rate being greater during inspiration than expiration. Interestingly, the expiratory limb of the flow-volume curve is remarkably similar in shape and size regardless of expiratory effort. This limitation of expiratory flow over most of the lung volume, independent of effort, is due to the dynamic compression of airways by intrathoracic pressure. As forced expiration proceeds from TLC to RV, airway resistance must be overcome, and the high pressures near the alveoli decline and reach atmospheric pressure at the mouth. The high intrapleural pressures generated collapse the downstream airways, and the pressure difference driving expiratory flow becomes alveolar minus intrapleural pressure instead of alveolar minus mouth pressure. In some highly-trained male endurance athletes, the mechanical limits for inspiratory and expiratory pressure development and flow generation can be reached during exercise [16]. There is also recent evidence to show that maximal expiratory flow is achieved during near-maximal exercise in endurance-trained female athletes

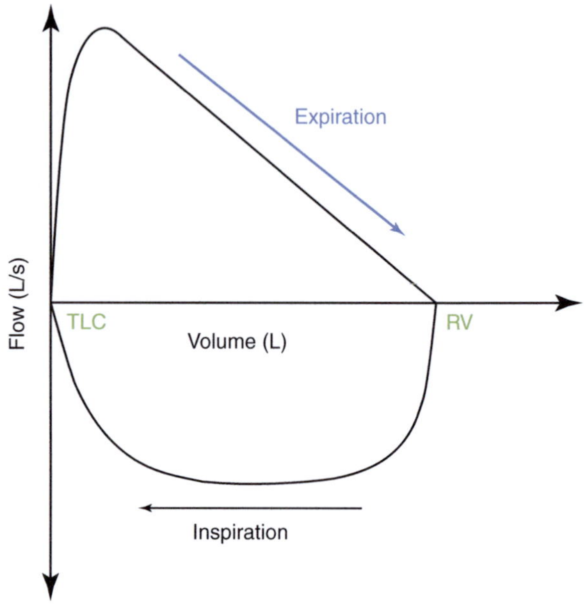

Fig. 1.1 Maximal flow-volume loop. *TLC* total lung capacity, *RV* residual volume

[17, 18]. The observation of expiratory flow limitation (EFL) during exercise in athletic humans suggests that the respiratory system is not without limits when the high ventilatory demands of strenuous exercise are superimposed.

1.4.3 Breathing Pattern

In addition to ensuring that blood gas homeostasis is maintained, the optimal breathing response also minimizes the amount of work required for a given $\dot{V}_A$. The amount of energy required to ventilate is referred to as the work of breathing (WOB). Important variables to consider for determining the mechanical WOB are tidal volume (V_T), breathing frequency (F_B), end-inspiratory lung volume (EILV), end-expiratory lung volume (EELV), flow rates, and duty cycle. The $\dot{V}_A$ is determined by the minute ventilation ($\dot{V}_E$) minus the dead space ventilation ($\dot{V}_D$). As the same level of $\dot{V}_E$ can be achieved with different combinations of V_T and F_B, a combination of these two variables must be established such that the increase in $\dot{V}_E$ is as efficient as possible and the $\dot{V}_D$ is minimized. At rest, inspiration occurs primarily due to diaphragm contraction, and expiration is passive. As $\dot{V}_E$ increases above rest, EELV falls due to recruitment of the expiratory abdominal muscles [19]. Recruitment of the abdominal muscles assists the inspiratory muscles in multiple ways. First, by reducing EELV, increases in V_T can occur on the linear portion of the respiratory system compliance curve [20]. Second, as EELV is reduced, the diaphragm is lengthened. Lengthening of the diaphragm allows it to function at a more optimal length for tension development, according to the length-tension relationship of the

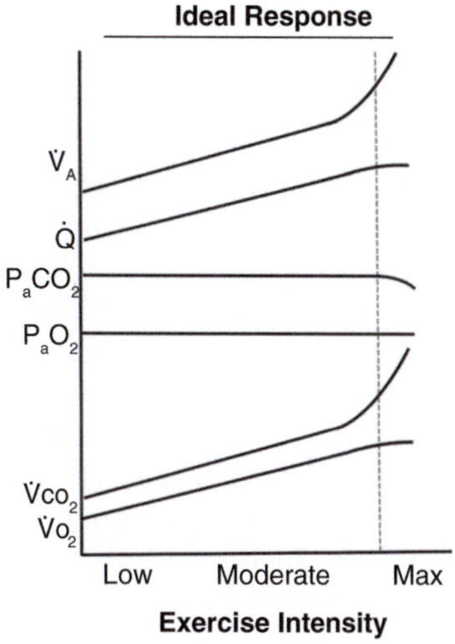

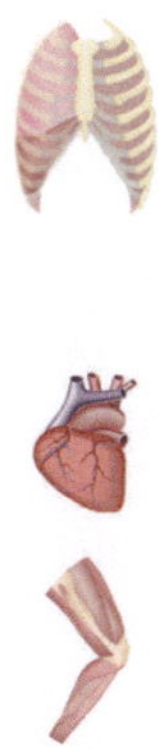

		Untrained Rest	Untrained Max	Trained Max
V_T	(L)	0.6	2.7	2.9
F_B	(breaths·min^{-1})	10	42	58
$\dot{V}_E$	(L·min^{-1})	6	113	168
$\dot{V}_A$	(L·min^{-1})	4	93	138
P_aCO_2	(mmHg)	39	31	38
P_aO_2	(mmHg)	91	92	70
HR	(beats·min^{-1})	77	190	190
SV	(mL)	65	105	180
$\dot{Q}$	(L·min^{-1})	5	20	34
$\dot{V}co_2$	(L·min^{-1})	0.19	3.3	6.04
$\dot{V}o_2$	(L·min^{-1})	0.24	3.0	5.25
pH		7.4	7.28	7.27

Fig. 1.2 Representative values for ventilation and pulmonary gas exchange at rest and during maximal exercise in a healthy, untrained young adult ($\dot{V}_{O_2\,max}$ 45 mL kg^{-1} min^{-1}) and at maximal exercise in an endurance-trained young adult ($\dot{V}_{O_2\,max}$ 70 mL kg^{-1} min^{-1}). V_T tidal volume, F_B breathing frequency, $\dot{V}_E$ minute ventilation, $\dot{V}_A$ alveolar ventilation, P_aCO_2 partial pressure of arterial carbon dioxide, P_aO_2 partial pressure of arterial oxygen, *HR* heart rate, *SV* stroke volume, $\dot{Q}$ cardiac output, $\dot{V}_{CO_2}$ carbon dioxide production, $\dot{V}_{O_2}$ oxygen uptake

muscle [21]. Third, contraction of the abdominal muscles allows storage of elastic energy in the abdominal and thoracic walls that can be used to make up a portion of the energy required for the subsequent inspiration [19]. During moderate increases in $\dot{V}_E$, there are increases in F_B and V_T (by reducing EELV and increasing EILV), whereas at higher levels of $\dot{V}_E$, increases in F_B dominate. A plateau in V_T ensures that breathing occurs along the linear portion of the respiratory compliance curve. Figure 1.2 provides representative values for ventilatory variables during rest and maximal exercise in a healthy adult and during maximal exercise in a trained endurance athlete.

1.5 Pulmonary Gas Exchange During Exercise

During exercise, $\dot{V}_A$ increases in proportion to $\dot{V}O_2$ and $\dot{V}CO_2$ according to the following alveolar gas equations:

$$P_ACO_2 = \left[\dot{V}CO_2 \div \dot{V}_A\right] \times K$$

$$P_AO_2 = P_IO_2 - \left[\dot{V}O_2 \div \dot{V}_A\right] \times K$$

where P_ACO_2 represents alveolar carbon dioxide partial pressure and is assumed to be approximately equal to arterial PCO_2; P_AO_2 is alveolar oxygen partial pressure; P_IO_2 is partial pressure of inspired oxygen; and K is a constant (0.863). The value of P_AO_2 and P_ACO_2 may be calculated using the above equations if VO_2 and VCO_2 are expressed in mL/min and $\dot{V}_A$ is expressed in L/min.

At sea level, with a P_B of 760, PO_2 is 160 mmHg. As air is inspired it is warmed to body temperature and completely saturated with water vapor. At body temperature, the water vapor exerts a pressure of 47 mmHg, and thus inspired air at sea level has a PO_2 of approximately 150 mmHg. By the time air reaches the alveoli, the PO_2 has fallen to about 100 mmHg due to continuous addition of O_2 to the alveoli by ventilation as well as the removal of O_2 by the pulmonary capillaries. Diffusion between the alveoli and the pulmonary capillaries occurs due to the alveolar-arterial PO_2 difference (A-aDO_2), which is determined by the lungs' ability to oxygenate blood that is returning to the lungs. As alveolar PO_2 can be calculated using the equation above and arterial PO_2 can be measured through blood gas analysis, the A-aDO_2 is easily determined. In a resting healthy human, A-aDO_2 ranges from 5 to 10 mmHg. When dynamic exercise is performed by healthy individuals, P_aO_2 and % oxyhemoglobin saturation remain at pre-exercise values across all exercise intensities, despite A-aDO_2 increasing 2–3 times above resting levels.

The ability of the healthy lung to maintain P_aO_2 despite the significant increase in metabolic rate associated with dynamic exercise is remarkable. To facilitate this, the surface area for gas exchange is enhanced by a three- to fourfold increase in pulmonary capillary blood volume. Increased pulmonary capillary blood volume, achieved via recruitment and distension of the pulmonary capillary beds, minimizes the reduction in red cell transit time (red cell transit time = pulmonary capillary blood volume/blood flow) that occurs with a large cardiac output. At maximal exercise, red cell transit time is approximately 0.5 s compared to 0.9 s at rest. Despite this significant reduction in time for the red blood cell to bind with oxygen, diffusion equilibrium occurs and P_aO_2 is maintained. An increased A-aDO_2 due to the hyperventilation associated with high-intensity exercise also protects P_aO_2.

There is evidence that all is not perfect with respect to arterial blood gas homeostasis during exercise in the endurance athlete. "Exercise-induced arterial hypoxemia" (EIAH) has been reported to variable degrees in highly trained endurance athletes and can be characterized by reductions in arterial oxygenation (SaO_2): mild 93–95%, moderate 88–93%, and severe <88%. An excessively wide A-aDO_2 (>25–30 mmHg) and inadequate compensatory hyperventilation ($PaCO_2 > 35$ Torr) commonly contribute to EIAH. A detailed treatment of this topic is beyond the scope of this chapter, and the interested reader is directed elsewhere [22].

1.5.1 Acid-Base Regulation

Carbon dioxide must be continuously removed from the body via the lungs, and its transport has a large effect on blood and tissue acid-base status. Venous blood transports CO_2 to the lungs in three forms: dissolved in plasma, as carbamino compounds, and as bicarbonate (HCO_3^-). CO_2 is approximately 24 times more soluble than O_2, and about 10% of the CO_2 transported to the lungs is in the dissolved form. Carbamino compounds, formed by the combination of CO_2 with plasma proteins and hemoglobin, transport 30% of CO_2 to the lungs. The remaining 60% of CO_2 is carried in the form of HCO_3^- formed by the hydration of CO_2 to carbonic acid (H_2CO_3) during the following reaction:

$$CO_2 + H_2O \leftrightarrow H_2CO_3 \leftrightarrow H^+ + HCO_3^-$$

The presence of the enzyme carbonic anhydrase increases the rate of the first reaction within the red blood cell, while dissociation of carbonic acid in the second reaction is fast without an enzyme. When the concentration of H^+ and HCO_3^- increases, HCO_3^- diffuses out of the red blood cell, but due to the cell membrane being relatively impermeable to cations, H^+ cannot. To balance the movement of negative ions out of the red blood cell, Cl^- ions diffuse into the cell and maintain electrical neutrality ("chloride shift"). Upon arrival at the lung, the above processes are reversed, and CO_2 is eliminated through alveolar ventilation. Compared with removal of acids that occurs via the kidney (about 100 milliequivalents per day), the lung excretes over 10,000 milliequivalents of carbonic acid per day. Therefore, by altering alveolar ventilation, the body has significant control over its acid-base status. During light exercise there is little change to the acid-base status of arterial blood, despite increases in mixed venous PCO_2 and reductions in pH. Metabolic acidosis occurs as exercise intensity is increased, resulting in arterial pH decreases and blood lactate increases. Increased exercise intensity also leads to hyperventilation and a concomitant decrease in P_aCO_2. The degree of hyperventilation varies between individuals, with P_aCO_2 decreasing as much as 15 mmHg in some people and as little as 2–3 mmHg in others. Figure 1.2 provides representative values for pulmonary gas exchange variables and pH during rest and maximal exercise in a healthy adult and during maximal exercise in a trained endurance athlete.

1.6 Cardiopulmonary Interactions

The human cardiovascular and respiratory systems are inseparably integrated such that manipulation of one system will have consequences for the other. This is particularly the case under conditions of dynamic exercise. For example, relative homeostasis of arterial blood gases is dependent upon the coordinated changes that link alveolar ventilation, cardiac output, blood flow, and tissue metabolism. As such, the interactions between the heart, vasculature, lungs, and muscles of respiration are an important consideration for our understanding of integrative human physiology. We have broadly classified cardiopulmonary interactions into two categories: (1) mechanical interactions and the respiratory pump and (2) respiratory muscle blood flow.

1.6.1 Mechanical Interactions and the Respiratory Muscle Pump

The heart and lungs occupy the intrathoracic space, and as such exercise-induced changes to volume and pressure must be considered for both organ systems. During inspiration under resting conditions, there is a corresponding negative intrathoracic pressure that is also "seen" by the heart. With inspiration, and the associated negative intrathoracic pressure, the pressure gradient across the walls of the heart increases and is thought to augment cardiac preload. This is accomplished by a reduction in right atrial pressure with inspiration which expands the pressure gradient for venous return. With expiration and the increase in intrathoracic pressure, there is an enhanced ventricular emptying which transiently increases stroke volume. Under conditions of systole at rest, the lowering of intrathoracic pressure with inspiration widens the transmural pressure gradient across the walls of the heart and hampers ventricular emptying. On the other hand, during expiration, the positive intrathoracic pressure augments ventricular emptying. Increases in abdominal pressure during diaphragmatic contraction can compress the inferior vena cava and reduce blood return from the lower limbs [23]. A further consideration is the amount of blood volume in the inferior vena cava. When blood volume in the inferior vena cava is high (preceding inspiration), a rise in abdominal pressure will force blood toward the heart, whereas when blood volume is low (end-inspiration), a small volume of blood will be forced upward [24].

Under conditions of dynamic exercise, it is well known that rhythmic contraction of the locomotor muscles pushes blood back to the heart and is commonly referred to as the "skeletal muscle pump." There is general agreement that the skeletal muscle pump is essential in maintaining venous return during exercise. How respiratory muscle contraction modulates venous return and cardiac output during dynamic exercise in humans is difficult to study owing to the invasiveness of the measures and anatomical complexities. However, there is evidence that there is a significant modulatory effect of respiration on venous return during cycling exercise [25]. Miller et al. [26] studied healthy humans who performed plantar flexion exercise while breathing with their diaphragm or with their rib cage. Blood flow in the working limb was assessed using Doppler ultrasound. There was significant within-breath modulation of femoral venous blood flow that persisted despite the presence of rhythmic calf contraction forcing blood centrally. These findings point to the fact that the cardiovascular system is susceptible to manipulation by respiratory pressures during exercise. Lastly, additional evidence to show that respiratory muscle pressure generation during exercise influences blood flow comes from studies that have used positive pressure mechanical ventilation during inspiration [27]. Here, the less negative intrathoracic pressure with mechanical ventilation resulted in significant decreases in stroke volume and cardiac output presumably because a less negative intrathoracic pressure limits cardiac preload. In sum, there appears to be good evidence to suggest that the pressures produced by the respiratory muscles during inspiration can contribute to the cardiac output and stroke volume responses to whole-body dynamic exercise.

1.6.2 Respiratory Muscle Blood Flow

Skeletal muscle blood flow and metabolism are closely matched during exercise [28]. Studying the coupling of blood flow and metabolism with respect to the respiratory musculature in humans is difficult for technical and anatomical reasons. As such, much of what is known about respiratory muscle blood flow during exercise comes from animal studies or indirect evidence in humans. For example, Manohor completed a number of complex studies examining respiratory muscle blood flow using radionuclide-labeled microspheres. Here it was shown that in maximally exercising equines the diaphragm and locomotor muscle vasculature achieved comparable and increased blood flow rates (>300 mL min^{-1} 100 g^{-1}) [29–31]. The blood flow to respiratory muscles was approximately 15–16% of the total cardiac output in the equine at maximum exercise [31], and qualitatively similar blood flow values have been reported in other species including rodents and dogs. In healthy, aerobically trained humans, a comparable fraction of cardiac output has been reported for respiratory muscles during maximal exercise [27]. These estimates of respiratory muscle blood flow are similar to those based on measures of "trunk and head" flow using dye dilution via catheterization of the subclavian and femoral veins [32]. It appears that blood flow to the respiratory musculature based on different experimental approaches and species is approximately 14–20% of cardiac output.

To sustain heavy exercise, the muscles of locomotion must generate significant force for a prolonged period. In order to ensure appropriate alveolar ventilation during exercise, the force output of the muscles of respiration is also substantial and must be considered. Evidence to support this concept comes from studies where high levels of ventilatory work are reduced during heavy exercise using a ventilator and blood flow to active limbs is increased [33]. Conversely, when the work of breathing is increased via inspiratory resistors, leg blood flow is decreased. The alterations in leg blood flow that result from the manipulation of ventilatory work imply a competitive relationship between locomotor and respiratory muscles for a finite cardiac output. The hypothesis here is that high levels of respiratory muscle work and the associated accumulation of metabolites in the respiratory muscles activate type III–IV phrenic afferents, which induce reflex increases in limb sympathetic vasoconstrictor activity or "respiratory muscle metaboreflex" [34]. Recently, Dominelli [35] reduced the WOB with a proportional assist ventilator in subjects cycling at near-maximal intensities and measured respiratory and leg muscle blood flow using near-infrared spectroscopy optodes placed over the sternocleidomastoid muscles, vastus lateralis, and vastus medialis coupled with a venous indocyanine green dye injection. Lowering the normally occurring WOB reduced sternocleidomastoid blood flow and increased leg blood flow (Fig. 1.3). These findings support the concept that respiratory muscle work significantly influences the distribution of blood flow to both respiratory and locomotor muscles. Unloading the respiratory muscles during high-intensity exercise prevents diaphragm fatigue [36] and attenuates locomotor muscle fatigue at end exercise [37] further emphasizing the integrated nature of respiratory influences on the physiology of whole-body exercise.

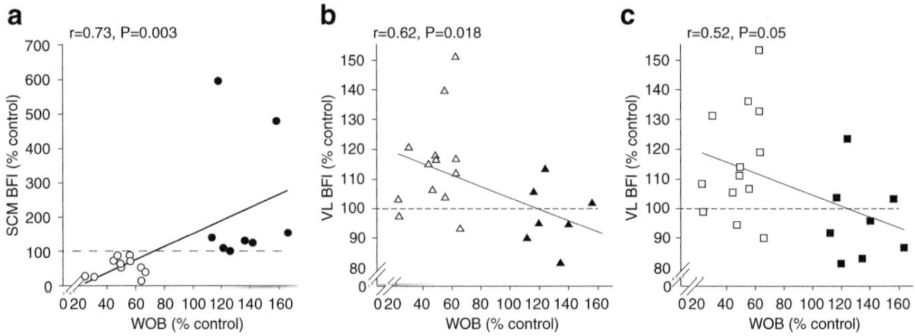

Fig. 1.3 Individual values for sternocleidomastoid (**a**), vastus lateralis (**b**), and vastus medialis (**c**) blood flow index and work of breathing. Open symbols represent trials while breathing on a proportional assist ventilator, and filled symbols represent trials breathing on a resistor. *BFI* blood flow index, *SCM* sternocleidomastoid, *VL* vastus lateralis, *VM* vastus medialis, *WOB* work of breathing. From Dominelli PB, Archiza B, Ramsook AR, Mitchell RA, Peters CM, Molgat-Seon Y et al. Effects of respiratory muscle work on respiratory and locomotor blood flow during exercise. Exp Physiol. 2017 102(11):1535–47

1.6.3 Mechanical Efficiency of Exercise Hyperpnea

As $\dot{V}_E$ increases, there is an associated increase in metabolic and circulatory costs of generating the mechanical WOB [38]. To determine how efficiently the respiratory muscles are able to produce the mechanical WOB, the oxygen cost of breathing ($\dot{V}O_{2RM}$) can be determined. The $\dot{V}O_{2RM}$ can be estimated by having subjects mimic their exercise breathing pattern at rest while measuring $\dot{V}O_2$. Respiratory muscle efficiency can then be calculated by dividing the measured $\dot{V}O_{2RM}$ by the ideal oxygen uptake needed to perform the measured WOB [39]. The ideal oxygen uptake can be calculated by converting the measured WOB into units of oxygen. When hyperpnea of exercise is mimicked at rest, with careful matching of operating lung volumes and pressure generation between exercise and mimicking trials, mechanical efficiency of the respiratory muscles has been estimated to range from 4 to 7% [39, 40]. The estimated mechanical efficiency in these studies is likely underestimated due to the difficulty in measuring all the work associated with breathing. Measuring WOB using an esophageal catheter does not account for the energy required to sustain muscle contraction [41], the work done to distort the chest wall at high ventilations [42], and the work associated with increased velocity of muscle shortening [43]. Given the substantial amount of work required to maintain $\dot{V}_E$ at levels 20–25 times greater than the rest, as well as the relatively low mechanical efficiency of exercise ventilation, the oxygen cost of sustained ventilation during endurance exercise is high. In healthy untrained subjects ($\dot{V}O_{2max}$, 40–50 mL kg^{-1} min^{-1}), the oxygen cost of ventilation is approximately 7–10% $\dot{V}O_{2max}$ and in highly trained subjects ($\dot{V}O_{2max}$, >60 mL kg^{-1} min^{-1}) is 13–16% of $\dot{V}O_{2max}$ [39, 40].

1.7 Summary

We have presented the case that in health, the human respiratory system is, in many ways, ideally structured and regulated to meet the demands of dynamic exercise. Increases in ventilation during high-intensity exercise serve to ensure appropriate oxygenation of arterial blood along with the elimination of metabolically produced CO_2 and pH regulation. This is accomplished via well-suited anatomical structures and a complex feedforward and feedback neural control system that requires coordinated contractions of the respiratory musculature. There are some notable exceptions to the concept of "ideal structure and function" where some otherwise healthy athletes (i.e., non-asthmatic or other cardiorespiratory diseases) exhibit exercise-induced arterial hypoxemia and fatigue of the diaphragm.

Key Points

- Pressure, volume, and flow rate demands placed on the pulmonary system are significantly increased during exercise.
- Feedforward (central command) and feedback from sensory receptors provide the stimuli for exercise hyperpnea.
- The capacity of the respiratory system is, in most cases, well-suited to meet the demands placed on it during dynamic exercise.

References

1. Weibel ER, Gomez DM. Architecture of the human lung. Science. 1962;137(3530):577–85.
2. Sauret V, Halson PM, Brown IW, Fleming JS, Bailey AG. Study of the three-dimensional geometry of the central conducting airways in man using computed tomographic (CT) images. J Anat. 2002;200(2):123–34.
3. Sheel AW, Romer LM. Ventilation and respiratory mechanics. Compr Physiol. 2012;2:1093–142.
4. Smith JC, Ellenberger HH, Ballanyi K, Richter DW, Feldman JL. Pre-Bötzinger complex: a brainstem region that may generate respiratory rhythm in mammals. Science. 1991;254(5032):726–9.
5. Hilaire G, Pásaro R. Genesis and control of the respiratory rhythm in adult mammals. Physiology. 2003;18(1):23–8.
6. Kaufman MP, Forster HV. Reflexes controlling circulatory, ventilatory and airway responses to exercise. In: Comprehensive Physiology, Supplement 29: handbook of physiology, exercise: regulation and integration of multiple systems. Hoboken, NJ: John Wiley & Sons, Inc.; 2011. p. 381–447.
7. Amann M. Significance of group III and IV muscle afferents for the endurance exercising human. Clin Exp Pharmacol Physiol. 2012;39(9):831–5.
8. Dempsey JA, Adams L, Ainsworth DM, Fregosi RF, Gallagher CG, Guz A, et al. Airway, lung, and respiratory muscle function during exercise. In: Terjung R, editor. Comprehensive Physiology. New York: Oxford University Press; 2011.
9. Van de Graaff WB. Thoracic influence on upper airway patency. J Appl Physiol. 1988;65(5):2124–31.

10. Connel DC, Fregosi RF. Influence of nasal airflow and resistance on nasal dilator muscle activities during exercise. J Appl Physiol. 1993;74(5):2529–36.
11. Williams JS, Janssen PL, Fuller DD, Fregosi RF. Influence of posture and breathing route on neural drive to upper airway dilator muscles during exercise. J Appl Physiol. 2000;89(2):590–8.
12. Wetter TJ, St. Croix CM, Pegelow DF, Sonetti DA, Dempsey JA. Effects of exhaustive endurance exercise on pulmonary gas exchange and airway function in women. J Appl Physiol. 2001;91(2):847–58.
13. Warren JB, Jennings SJ, Clark TJH. Effect of adrenergic and vagal blockade on the normal human airway response to exercise. Clin Sci. 1984;66(1):79–85.
14. Strohl KP, Norcia MP, Wolin AD, Haxhiu MA, van Lunteren E, Deal EC. Nasal and tracheal responses to chemical and somatic afferent stimulation in anesthetized cats. J Appl Physiol. 1988;65(2):870–7.
15. Fredberg JJ, Inouye D, Miller B, Nathan M, Jafari S, Helioui Raboudi S, et al. Airway smooth muscle, tidal stretches, and dynamically determined contractile states. Am J Respir Crit Care Med. 1997;156(6):1752–9.
16. Johnson BD, Saupe KW, Dempsey JA. Mechanical constraints on exercise hyperpnea in endurance athletes. J Appl Physiol. 1992;73(3):874–86.
17. McClaran SR, Harms CA, Pegelow DF, Dempsey JA. Smaller lungs in women affect exercise hyperpnea. J Appl Physiol. 1998;84(6):1872–81.
18. Dominelli PB, Guenette JA, Wilkie SS, Foster GE, Sheel AW. Determinants of expiratory flow limitation in healthy women during exercise. Med Sci Sport Exerc. 2011;25:1666–74.
19. Henke KG, Sharratt M, Pegelow D, Dempsey JA. Regulation of end-expiratory lung volume during exercise. J Appl Physiol. 1988;64(1):135–46.
20. Stubbing DG, Pengelly LD, Morse JL, Jones NL. Pulmonary mechanics during exercise in normal males. J Appl Physiol. 1980;49(3):506–10.
21. Smith J, Bellemare F. Effect of lung volume on in vivo contraction characteristics of human diaphragm. J Appl Physiol. 1987;62(5):1893–900.
22. Dempsey JA, Wagner PD. Exercise-induced arterial hypoxemia. J Appl Physiol. 1999;87(6):1997–2006.
23. Willeput R, Rondeux C, De Troyer A. Breathing affects venous return from legs in humans. J Appl Physiol. 1984;57(4):971–6.
24. Takata M, Wise RA, Robotham JL. Effects of abdominal pressure on venous return: abdominal vascular zone conditions. J Appl Physiol. 1990;69(6):1961–72.
25. Wexler L, Bergel DH, Gabe IT, Makin GS, Mills CJ. Velocity of blood flow in normal human venae cavae. Circ Res. 1968;23(3):349–59.
26. Miller JD, Pegelow DF, Jacques AJ, Dempsey JA. Effects of augmented respiratory muscle pressure production on locomotor limb venous return during calf contraction exercise. J Appl Physiol. 2005;99(5):1802–15.
27. Harms CA, Wetter TJ, McClaran SR, Pegelow DF, Nickele GA, Nelson WB, et al. Effects of respiratory muscle work on cardiac output and its distribution during maximal exercise. J Appl Physiol. 1998;85(2):609–18.
28. Andersen P, Saltin B. Maximal perfusion of skeletal muscle in man. J Physiol. 1985;366:233–49.
29. Manohar M. Blood flow to the respiratory and limb muscles and to abdominal organs during maximal exertion in ponies. J Physiol. 1986;377:25–35.
30. Manohar M. Costal vs. crural diaphragmatic blood flow during submaximal and near-maximal exercise in ponies. J Appl Physiol. 1988;65(4):1514–9.
31. Manohar M. Inspiratory and expiratory muscle perfusion in maximally exercised ponies. J Appl Physiol. 1990;68(2):544–8.
32. Calbet JAL, Gonzalez-Alonso J, Helge JW, Søndergaard H, Munch-Andersen T, Boushel R, et al. Cardiac output and leg and arm blood flow during incremental exercise to exhaustion on the cycle ergometer. J Appl Physiol. 2007;103(3):969–78.
33. Harms CA, Babcock MA, McClaran SR, Pegelow DF, Nickele GA, Nelson WB, et al. Respiratory muscle work compromises leg blood flow during maximal exercise. J Appl Physiol. 1997;82(5):1573–83.

34. Dempsey JA, Romer L, Rodman J, Miller J, Smith C. Consequences of exercise-induced respiratory muscle work. Respir Physiol Neurobiol. 2006;151(2):242–50.
35. Dominelli PB, Archiza B, Ramsook AR, Mitchell RA, Peters CM, Molgat-Seon Y, et al. Effects of respiratory muscle work on respiratory and locomotor blood flow during exercise. Exp Physiol. 2017;102(11):1535–47.
36. Babcock MA, Pegelow DF, Harms CA, Dempsey JA. Effects of respiratory muscle unloading on exercise-induced diaphragm fatigue. J Appl Physiol. 2002;93(1):201–6.
37. Romer LM, Lovering AT, Haverkamp HC, Pegelow DF, Dempsey JA. Effect of inspiratory muscle work on peripheral fatigue of locomotor muscles in healthy humans. J Physiol. 2006;571(2):425–39.
38. Otis AB. The work of breathing. Physiol Rev. 1954;34(3):449–58.
39. Aaron EA, Seow KC, Johnson BD, Dempsey JA. Oxygen cost of exercise hyperpnea: implications for performance. J Appl Physiol. 1992;72(5):1818–25.
40. Dominelli PB, Render JN, Molgat-Seon Y, Foster GE, Romer LM, Sheel AW. Oxygen cost of exercise hyperpnoea is greater in women compared with men. J Physiol. 2015;593(8):1965–79.
41. Field S, Sanci S, Grassino A. Respiratory muscle oxygen consumption estimated by the diaphragm pressure-time index. J Appl Physiol. 1984;57(1):44–51.
42. Grimby G, Bunn J, Mead J. Relative contribution of rib cage and abdomen to ventilation during exercise. J Appl Physiol. 1968;24(2):159–66.
43. McCool FD, McCann DR, Leith DE, Hoppin FG. Pressure-flow effects on endurance of inspiratory muscles. J Appl Physiol. 1986;60(1):299–303.

Part II

Respiratory Assessment and Physical Activity

Medical History, Questionnaires and Physical Examination

2

Paolo Onorati and Giuseppe Fiorenzano

Abstract

Negative cardiovascular events represent the main risks associated with exercise. Also in healthy individuals, such risks increase during high-intensity exercise compared to rest, but still the absolute risk of a cardiac event remains low. Of course, high-intensity exercise increases significantly the risk of such events in subjects with cardiovascular diseases. Instead, respiratory diseases are rarely a contraindication for physical or sport activities but may condition their practice and therefore should be diagnosed. A lack of consensus still exists on the extent of medical investigations required before prescribing any exercise programme, even if it is of high intensity. Medical history and physical examination are mandatory to detect subjects at moderate to high risk of cardiovascular events. Questionnaires can only contribute marginally to this evaluation process, whereas they surely represent a useful tool for measuring physical activity. The medical history should be sufficiently detailed and tailored to include past and current informations, including those regarding daily physical activity. For this last aspect, questionnaires, besides being inexpensive and easy to apply, are also sensitive and suitable to detect relevant characteristics of patients' daily life physical activities and related limitations. However, there is still an insufficient knowledge about reliability, validity and sensitivity of such questionnaires for older

P. Onorati (✉)
U.O.C. Medicina Interna, Ospedale Civile di Alghero, ASSL-Sassari, ATS Sardegna, Alghero, Sassari, Italy

Department of Public Health and Infectious Diseases, Sapienza University of Rome, Rome, Italy

Internal Medicine and Pulmonary Division, Adult Cystic Fibrosis Program, Policlinico Umberto I Hospital, Rome, Italy

G. Fiorenzano
SSD Pneumologia, Azienda Ospedaliera Universitaria S. Maria, Terni, Italy

A. Cogo et al. (eds.), *Exercise and Sports Pulmonology*,
https://doi.org/10.1007/978-3-030-05258-4_2

adults or patients with chronic diseases. The physical examination should allow to detect signs that in combination with the medical history and characteristics of symptoms may be suggestive of some specific disease.

2.1 Introduction

The aim of the present chapter is to provide informations on the evaluation process prior to any type of exercise prescription, from physical fitness to clinical exercise testing. By definition physical fitness corresponds to a set of individual characteristics, either health- or skill-related, that influence the ability to perform a certain type or intensity of physical activity.

Risks associated with exercise are usually of cardiovascular nature (e.g. myocardial infarction or sudden cardiac arrest), which usually do not occur in healthy individuals while performing light- to moderate-intensity physical activity [1, 2]; such risks increase during high-intensity exercise compared to rest, but still the absolute risk of a cardiac event remains low. Of course, high-intensity exercise increases the risk of such events in subjects with cardiovascular diseases [3–5]. Chronic respiratory diseases are rarely considered as absolute contraindications for physical or sport activities but may be responsible for significant limitations to their execution, and, therefore, they need to be diagnosed and treated (e.g. asthma and/or allergic diseases especially in young subjects, chronic obstructive pulmonary disease in older ones particularly active and previous smokers) [6, 7]. Medical history and physical examination are mandatory to detect or to better characterize individuals at moderate to high risk of cardiovascular events. Questionnaires on physical activity can partly contribute to this evaluation process. However, there is still lack of consensus on the extent of medical investigations (e.g. stress testing) required before prescribing any exercise programme, even if it is of high intensity. The ATS/ERS task force for standardization of spirometry suggests the execution of a spirometry for the assessment of health status before starting strenuous physical activity programmes [8]. It is reasonable, as also considered by the American College of Cardiology/American Heart Association (ACC/AHA) [9], to prescribe exercise stress testing in individuals with increased risk of cardiovascular diseases (CVD). In Table 2.1 are listed the American College of Sports Medicine (ACSM) recommendations [10] for exercise testing prior to initiating physical activity. These considerations obviously do not concern the evaluation of athletes or subjects participating in competitive sports activities, for whom stress testing is required.

2.2 Medical History

The medical history should be sufficiently detailed and tailored to include past and current information (see Box 2.1). The following are the components of medical history that should be appropriately included:

Table 2.1 ACSM recommendations for exercise testing prior to initiating physical activity

1. New possible or unstable symptoms of cardiovascular disease.
2. Diabetes mellitus and at least one of the following factors:
(a) Age > 35 years.
(b) Type 2 diabetes mellitus >10-year duration.
(c) Type 1 diabetes mellitus >15-year duration.
(d) Hypertension with systolic blood pressure > 140 or diastolic >90 mmHg.
(e) Hypercholesterolaemia with total cholesterol >240 mg/dL.
(f) Smoking.
(g) Presence of microvascular disease.
(h) Peripheral artery disease.
(i) Autonomic neuropathy.
(j) Family history of CAD in first-degree relative <60 years.
3. End-stage renal disease.
4. Symptomatic or diagnosed chronic pulmonary disease (e.g. COPD, asthma, interstitial lung disease, cystic fibrosis).

1. Risk factors for cardiovascular and pulmonary diseases: hypertension, smoking history, alcoholic habit, professional exposure, diabetes, dyslipidaemia, obesity, metabolic syndrome.
2. Past medical history.
   (a) Cardiovascular diseases: coronary artery disease (CAD) (myocardial infarction, acute coronary syndromes), heart failure, valvular dysfunction, peripheral vascular disease, deep vein thrombosis.
   (b) Previous cardiovascular interventions: percutaneous coronary interventions (angioplasty and stenting), coronary artery bypass surgery, valvular surgery, pacemaker and cardioverter defibrillator implant, ablation interventions for arrhythmias, cardiac transplant.
   (c) Pulmonary diseases: chronic obstructive pulmonary disease (COPD), asthma and allergic diseases, interstitial lung diseases, cystic fibrosis, pneumothorax, pleuritis, previous thoracic procedures.
   (d) Cerebrovascular diseases: stroke, transient ischaemic attacks, epilepsy.
   (e) Blood dyscrasias.
   (f) Musculoskeletal disorders: congenital or acquired myopathies, osteoporosis, osteoarthritis, rheumatoid arthritis, spondyloarthritis, degenerative disc disease, orthopaedic interventions for bone fractures, articular replacement.
   (g) Cancer.
3. Family history of cardiovascular, pulmonary (asthma, emphysema, cystic fibrosis, atopy) and metabolic diseases, sudden death.
4. Social history: in addition to tobacco and alcohol also drug use and/or addiction, home/work exposure.
5. Exercise history and physical activity, which means any bodily movement produced by skeletal muscles that results in energy expenditure [11]. In clinical setting, it can be assessed by the use of diaries and questionnaires.

Furthermore, any kind of symptoms reported during or after physical exercise (chest pain, cough, palpitation, dyspnoea) should also be included. For physical and sport activity in special environments, medical history should include symptoms related to specific pathologies like high-altitude illness (headache, dizziness, nausea, dyspnoea) [12] or diving-related illness (decompression sickness, diving-induced pulmonary oedema) [13, 14].

2.3 Questionnaires

Physical activity is considered an important index for the evaluation of the efficacy either of therapeutic interventions in clinical trials or of public health interventions. There is a significant evidence demonstrating the importance of daily life physical activity in the management and prevention of many chronic diseases [15–19]. Many objective (e.g. pedometers, accelerometers, GPS technology) and subjective assessments are available for the evaluation of physical activity. Evaluating daily life physical activity by the use of questionnaires and/or diaries has the main advantage of being inexpensive and easy to apply. Patient-reported outcome (PRO) questionnaires represent the most common subjective tool of assessment, which consists in self-report measures of own health condition directly expressed by the patient without any external interpretation. Even though an association exists between the tested physical activity questionnaires and health/functioning variables, the formers are potentially more sensitive and suitable to capture relevant characteristics of patient daily life physical activities and related limitations. To this end, it is important that the format of the questionnaire is correctly structured and characterized by questions and answers that are easily comprehensive for the patient, so as to better orient the therapeutic strategy in a more appropriate and patient-centred manner.

It is known that many factors may influence the accuracy of these techniques or induce bias in the assessment, such as:

(a) Individual characteristics such as cognitive capacity, age and cultural and social factors [20–22].
(b) The accuracy in the perception of certain type of activities, especially those of light intensity (e.g. self-care, home management) [23].
(c) The timing of information recall, which may differ from periods of hours to the entire lifetime [24, 25] and may cause difficulties, especially in elderly subjects, when recalling over long period of time; in fact, reliability of information recalled may decrease in relation to the length of the period examined [26].
(d) The structure and the design of a questionnaire, either in terms of comprehensibility or simplicity, that can provide better reliability and prevent subjects from being demotivated or feel confused by long and complicated questionnaires [26, 27], either in terms of response options, where it has been documented a higher amount of self-reported physical activity by the use of interval response questions compared to open options [28].

It should be also considered the reduced accuracy of energy expenditure estimation by the use of tables of energy cost of different physical activities, which may vary significantly between subjects due to different factors, such as body weight or movement skill efficiency; moreover, there's still a lack of information on the energy cost of many daily life activities [29, 30].

Although there have been several reviews on physical activity PRO questionnaires in recent years, the knowledge about reliability, validity and sensitivity of questionnaires for older adults or patients with chronic diseases (e.g. COPD) [31] is still insufficient, and more high-quality validation studies are needed.

Reliability consists in the ability of an instrument to yield correlated results when applied to the same population under similar conditions on at least two successive occasions [32]. Low reliability may be found when reporting low-intensity activities and/or in long test–retest intervals, as well as when a questionnaire is submitted by different interviewers [33, 34]. Test–retest reliability has been investigated in several questionnaires (e.g. the Physical Activity Scale in the Elderly or PASE, the Minnesota Leisure Time Physical Activity Questionnaire, Zutphen Physical Activity Questionnaire or ZPAC, Baecke's questionnaire) when administered to the general elderly population [35–38], while there is a lack of information on this issue in patients with chronic diseases such as COPD. It must be noted that test–retest reliability may be also negatively influenced by the day-to-day variability of subjects' daily physical activity, which, in turn, may be caused by different factors such as age or employment status [39].

Validity is the assessment of whether an instrument measures what is actually supposed to measure [32]. It should be indicated by the correspondence of the instrument's results to the criterion or "gold standard" method, such as objective measurements of physical activity (e.g. motion sensors, oxygen consumption measurements) or other validated questionnaires. First of all, cultural confounding factors should be taken into account when utilizing translated questionnaires [40]. Many evidences in the literature have shown that questionnaires correlated well with the criterion methods mostly for high-intensity physical activity and less for those of light or moderate intensity [27]; this therefore may reduce the validity of these questionnaires if applied to patient populations that are more compromised in terms of physical activity.

Sensitivity, also intended as responsiveness, refers to the capacity of an instrument to detect statistically significant changes over time that are expected to occur [41] and is closely linked to the reliability and validity of the instrument itself. Questionnaire sensitivity to interventions (e.g. rehabilitation programmes) has been documented in elderly populations and COPD patients for some aspects of daily physical activity (e.g. functional status, quality of life) [42–44].

Different systematic reviews on subjective physical activity monitoring have been published, which considered and analysed questionnaires developed for different categorized populations such as the elderly, patients with chronic respiratory disorders and in particular with COPD, patients with other specified diseases (e.g. cardiovascular, musculoskeletal and rheumatoid disorders) and also patients with unspecified chronic diseases or disability [31, 45–47]. As pointed out in these reviews, questionnaires should not be focused only on physical activity alone but also on different physical activity domains, such as daily life activity, general activity and mobility and work or social or leisure time activities, and should consider related symptoms. The format of answer options may be also very heterogeneous, such as numerical, categorical (e.g. yes/no) or visual analogue scales. Therefore, it may be difficult to decide, among different options, which is the most appropriate for a specific study population or type of questionnaire used.

It should be considered also that PRO questionnaires alone are not sufficient to capture all the significant aspects of the physical activity in patients with chronic diseases, such as differentiating specific patterns of physical activity or highlighting even minimal changes on a day-to-day basis. Probably the combination of both objective and subjective measurements should allow a more accurate evaluation of physical activity in chronic diseases, as pointed out in a recent study on COPD population conducted within the framework of the European Union Innovative Medicines Initiative PROactive project (www.proactivecopd.com) [48]. In this multicentre study, authors conducted a 6-week, randomized, two-way crossover assessment of physical activity in a COPD, combining informations obtained from two versions of PRO questionnaires (characterized by a daily and a 7-day recall assessment, respectively) and two activity monitors (accelerometers). They concluded that items generated from subjective experiences in combination with measurements of an activity monitor should provide a more comprehensive assessment of all relevant dimensions of physical activity in patients with COPD [49].

Questionnaires have been also created for the assessment and management of chronic respiratory disorders, such as asthma and COPD. In the management of patients with asthma, different questionnaires, either categorical (e.g. the Royal College of Physicians three questions) [50] or numerical (e.g. Asthma Control Questionnaire, Asthma Control Test) [51, 52], have been developed, in particular for the evaluation of symptom control, an aspect of considerable importance in the therapeutic decision-making process. As well as in the assessment and management of patients with COPD, formal symptomatic assessment tools based on questionnaires have been required. Between these, the Modified British Medical Research Council (mMRC) questionnaire has been considered adequate for assessment of symptoms (e.g. dyspnoea), as it relates with measures of health status and future mortality risk [53–56]. More comprehensive disease-specific health status questionnaires have been also developed, some probably too complex to use in routine practice (e.g. St. George's Respiratory Questionnaire) [57], while others are simpler but still suitable [58, 59].

Other types of questionnaires have been conceived and validated as screening and/or evaluation tools of the incidence and/or prevalence of particular diseases in specific study populations (e.g. athletes) [60].

2.4 History of Symptoms and Physical Examination

The physical examination allows to detect signs that in combination with the history and characteristics of symptoms may be suggestive, although often not specific, of cardiovascular, pulmonary or other (e.g. neurological, metabolic) diseases [61, 62].

Chest discomfort is one of the principal manifestations of cardiac disease and in particular coronary artery disease (CAD) but is not specific, because it may occur also in pulmonary and other diseases (e.g. neuromuscular). Features suggesting a cardiac ischaemic origin are the location, usually anteriorly in the midthorax or substernal, with irradiation to the neck and/or to the jaw and/or to ulnar side of the left forearm or both arms; the character is usually constricting or squeezing and often is elicited by some provoking factors such as exertion or other forms of stress. An ischaemic origin is less likely if the symptom is provoked by specific body motion or if is aggravated by respiration.

Palpitations are often described as "being aware of a rapid or forceful heart beating" and may be provoked by different disorders of heart rhythm, both bradycardic and tachycardic (tachycardia, ectopic beats), or by valvular impairments (e.g. valvular regurgitation). Palpitations may occur also in the presence of gas exchange (e.g. hypoxaemia) or circulatory (e.g. pulmonary hypertension) abnormalities present also in other diseases (e.g. pulmonary, neuromuscular, obesity, respiratory sleep disorders).

Cough, productive or not, is very common in respiratory diseases but may be present also in others. Recently onset or acute cough may be caused by viral or bacterial respiratory infections. Chronic cough in smokers is usually due to COPD. Otherwise, rhinosinusitis or gastro-oesophageal reflux may be involved. The cough may be a manifestation of bronchial hyperreactivity, especially in subjects with a family history of allergy and/or symptoms of rhinitis. Haemoptysis may be seen during an acute respiratory infection and may require clinical investigations to exclude bronchiectasis, tuberculosis or lung cancer.

Short of breath or dyspnoea is usually described as uncomfortable breathing that may occur at rest or during exertion and is characteristic either of cardiac or pulmonary diseases. A cardiac origin (left ventricular dysfunction) may be suspected when it occurs at rest in recumbent position, especially with paroxysmal nocturnal character, and is relieved by sitting or removing from the supine position. Nocturnal dyspnoea may be experienced also by patients with COPD but is usually associated with bronchial secretions and may be relieved by their clearance and/or the prompt use of medical devices (e.g. rapid response bronchodilators). Unexpected or abnormal exertional dyspnoea suggests a cardiopulmonary disease. If the dyspnoea occurs during or, most frequently, after exercise, an exercise-induced bronchospasm (EIB) may be suspected [63, 64]. A rare but potentially fatal syndrome is exercise-induced anaphylaxis (EIAn) [65], which is frequently related to food ingestion.

Syncope can be defined as a transient and complete loss of consciousness with a rapid onset, associated with inability to maintain postural tone, with rapid and spontaneous recovery. The presumed mechanism is cerebral hypoperfusion. Clinical features of other nonsyncope causes of loss of consciousness must be absent (e.g. seizure, pseudosyncope). Cardiac syncope may be caused by bradycardia, tachycardia or hypotension due to low cardiac index, blood flow obstruction, vasodilatation or acute vascular dissection [66, 67].

Intermittent claudication is characterized by muscle pain (often described as a cramp) in the legs during activity (e.g. walking), and it resolves within 1–2 min after stopping the activity. It is the consequence of an inadequate circulatory supply of blood and oxygen to the exercising muscles, usually secondary to atherosclerosis; this explains why CAD has a high prevalence in subjects with this disorder. Diabetes represent a major risk factor for this condition.

Inspection during physical examination may reveal: chest wall abnormalities are very important for the physiological adaptations to exercise. Kyphoscoliosis may cause functional restrictive respiratory defect and is characterized by an increased work of breathing. Pectus excavatum and pectus carinatum may be associated with functional respiratory abnormalities and also with cardiovascular malformations (Marfan's syndrome).

The digital clubbing is associated with various pulmonary diseases: bronchiectasis, idiopathic pulmonary fibrosis (IPF) and lung cancer.

Jugular vein turgor, assessed with the patient positioned supine with the head-neck-thorax axis at 45°, is expressive of increased jugular vein pressure, which may be caused by different cardiocirculatory abnormalities (e.g. right ventricular failure, tricuspid stenosis or regurgitation, pericardial effusion or constrictive pericarditis, superior vena cava obstruction, fluid overload, hyperdynamic circulation).

Soft tissue oedema of the ankles and/or legs and/or feet when bilateral is characteristic of congestive heart failure or bilateral chronic venous insufficiency, especially if it worsens in the evening. It may occur in patients with COPD, in the presence of gas exchange abnormalities (e.g. persistent hypoxaemia and/or hypercapnia) or right ventricular dysfunction (e.g. "cor pulmonale"). Generalized oedema or anasarca can be observed not only in severe chronic heart failure but also in chronic hepatic (e.g. cirrhosis) or renal (e.g. nephrotic syndrome) disorders.

Heart auscultation during physical examination may reveal extra sounds or heart murmurs known or found, which may represent valvular and/or cardiovascular disease. Among these, hypertrophic cardiomyopathy and aortic stenosis must be ruled out because they represent the more common causes of sudden cardiac death during exercise. Auscultation should be performed with the patient in the supine position, with the diaphragm of stethoscope at right second interspace at the sternal border (aortic area), left second interspace (pulmonic area) and then left third, fourth and fifth interspaces (tricuspid area) and at the apex fifth interspace at midclavicular line (mitral area). Intensity, rhythm and splitting of sound on each event of the cardiac cycle should be noted. Positioning the patient over onto left side and auscultating with the bell of stethoscope at the apex (mitral area) can help to detect some extra sounds or murmurs (low-pitched sounds of S3 and S4 and murmur of mitral stenosis; see below), or asking patient to sit up, lean forward and hold breath in exhalation, listening with diaphragm of stethoscope along left sternal border and at the apex and pausing periodically for patient to breathe can help to accentuate aortic murmurs or to detect pericardial friction rub. The heart first sound (S1/"lub") may be accentuated (e.g. tachycardia, anaemia, hyperthyroidism, mitral stenosis), diminished (e.g. mitral regurgitation, congestive heart failure) or varying (e.g. complete heart block, atrial fibrillation); an abnormal splitting may be due to right bundle branch block or premature ventricular contractions. The heart second sound (S2/"dub") may be physiologically split when it disappears on expiration, whereas it is pathological when it occurs during expiration (e.g. atrial septal defect, heart block) or is persistent (e.g. aortic stenosis, pulmonary hypertension).

Extra sounds in systole may be represented by high-pitched, sharp clicks that may be heard early (shortly after S1) at the base and apex (e.g. dilated aorta, valve disease or congenital stenosis) or in the second and third interspace (e.g. dilatation of pulmonary artery, pulmonary hypertension, pulmonic stenosis). A systolic click heard in mid or late systolic (before S2) at the apex or left lower sternal border is usually due to mitral valve prolapse. Extra sound in diastole may be represented by (a) an opening snap heard after the A2 (aortic) component of the second heart sound (S2), which correlates to the forceful opening of the mitral valve (e.g. mitral stenosis); (b) pathological third sound S3 (ventricular gallop) which occurs at the beginning of the middle third of diastole, approximately 0.12 to 0.18 s after S2 and may indicate serious problems like heart failure; and (c) fourth heart sound or S_4 (atrial gallop) which is an extra heart sound that may occur just after atrial contraction and immediately before the systolic S_1 and is caused by the atria contracting forcefully in an effort to overcome an abnormally stiff or hypertrophic ventricle.

Heart murmurs can be midsystolic (e.g. pulmonic stenosis, aortic stenosis, hypertrophic cardiomyopathy), pansystolic (e.g. mitral regurgitation, tricuspid regurgitation, ventricular septal defect) and diastolic (e.g. aortic regurgitation, mitral stenosis).

Lung auscultation may reveal adventitious breath sounds, such as:

(a) Fine crackles/rales, intermittent, late in inspiration and/or expiration, nonmusical, brief and high-pitched (as, e.g. pneumonia, early heart failure, bronchitis) or coarse crackles/rales intermittent, early or late in inspiration and/or expiration and louder (as, e.g. fibrosis, bronchiectasis).
(b) Wheezes longer, musical, intermittent, high pitch, hissing or shrill that are expression of narrowed airways, predominant during expiration (e.g. asthma, COPD, bronchitis if cleared with cough) or persistent and localized during inspiration and/or expiration (e.g. tumour/foreign body).
(c) Stridor is a predominant inspiratory wheeze, louder in the neck than the chest, which may result from a partial obstruction of the larynx or trachea (e.g. foreign body or croup).
(d) Pleural rub is creaking and usually confined to a small area and in both phases, in- and expiratory, which is the expression of inflamed, roughened surfaces grating against each other (e.g. pleurisy).
(e) Rhonchi are longer, musical and intermittent with lower pitch than wheezes and snoring quality and may be due to secretions in large airways (e.g. bronchitis, often cleared with cough).

Key Points

- Risks associated with exercise are usually of cardiovascular nature and may increase significantly during high-intensity exercise in subjects with cardiovascular diseases or correlated risk factors. Lack of consensus still exists on the extent of medical investigations required before prescribing any exercise programme.
- The respiratory diseases rarely contraindicate physical exercise and sport but may condition their practice and therefore should be detected and carefully evaluated.
- Questionnaires are useful to capture relevant characteristics of patients' daily life physical activities and related limitations, but still there is a lack of knowledge about their reliability, validity and sensitivity in older adults or patients with chronic diseases.
- A careful physical examination may allow to detect signs that in combination with the medical history and characteristics of symptoms may be indicative of some specific disease.

Box 2.1: A Model of Medical History Collection

(A) **General Informations**

First and last name:
Home address:
E-mail:
Phone: Home:_________ Work: __________Mobile:______________
Occupation:
Date of birth:________________Age__________
Gender: ☐ Male ☐ Female

(B) **Past medical history**

1. Childhood illnesses with residual abnormalities or health concerns for the future? (i.e. Polio with isolated weaknesses; Rheumatic Fever with heart valve damage, etc.)
   ☐ Yes ☐ No
   If yes, please explain:
   ____
2. Adulthood illnesses:

Heart Disease	☐ Yes ☐ No	High Blood Pressure	☐ Yes ☐ No
Diabetes	☐ Yes ☐ No	Asthma	☐ Yes ☐ No
Emphysema/COPD	☐ Yes ☐ No	Other Lung Disease	☐ Yes ☐ No
Kidney disease	☐ Yes ☐ No	Lung Cancer	☐ Yes ☐ No
Other Cancer	☐ Yes ☐ No	High Cholesterol	☐ Yes ☐ No

Anaemia ☐ Yes ☐ No Other blood dyscrasia ☐ Yes ☐ No
Stroke or other central Nervous System injuries ☐ Yes ☐ No
Muscular and/or osteoarticular disease ☐ Yes ☐ No
Unusual Infections ☐ Yes ☐ No Other Illnesses ☐ Yes ☐ No
If yes, further details below:

____

3. Hospitalization for anything other than surgery?
   ☐ Yes ☐ No
   If yes, for what, and when?

   ____

4. Surgical procedures in the past ☐ Yes ☐ No
   If yes, which type and when?

   ____

5. Injury in the past with any compromise of function?
   ☐ Yes ☐ No
   If yes, details:

   ____

6. Any specialized diagnostic procedures in the past? (i.e. heart catheterization, CAT or MRI scans, ergometric or stress test studies, etc.)
   ☐ Yes ☐ No.
   If yes, provide details:

   ____

7. List of medications and dosages

   ____

8. Vaccinations:
   ☐ Pneumovax ☐ Influenza ☐ Hepatitis A / B series ☐ others
   Date:____
9. History of allergies?
   Inhaled ☐ Yes ☐ No Food ☐ Yes ☐ No
   Contact ☐ Yes ☐ No Drug ☐ Yes ☐ No
   Iodine or Intravenous Contrast Dye? ☐ Yes ☐ No
   If yes, provide details below:

   ____

(C) **Family history**

Is your father living? ☐ Yes ☐ No Age (or age at death)?_____
Diseases:
☐ Heart Disease ☐ Diabetes ☐ Lung Disease/Emphysema/COPD
☐ Cancer ☐ High Cholesterol ☐ High Blood Pressure
☐ Serious Infections ☐ Other Illnesses
Please provide details:

____

Is your mother living? ☐ Yes ☐ No Age (or age at death)?_____
Diseases:

☐ Heart Disease ☐ Diabetes ☐ Lung Disease/Emphysema/COPD
☐ Cancer ☐ High Cholesterol ☐ High Blood Pressure
☐ Serious Infections ☐ Other Illnesses
Please provide details:
____

Are any siblings deceased? ☐ Yes ☐ No Age (or age at death)?_____
Diseases:
☐ Heart Disease ☐ Diabetes ☐ Lung Disease/Emphysema/COPD
☐ Cancer ☐ High Cholesterol ☐ High Blood Pressure
☐ Serious Infections ☐ Other Illnesses
Please provide details:
____

(D) **Social history**

Tobacco use: ☐Yes ☐No if yes which products:_______________
current smoker ☐Yes ☐ No ex-smoker ☐Yes ☐ No
How many years______ years n° Cigarettes/Cigar/other per day____
Packs of cigarette/year____
Alcohol use: ☐Yes ☐No, if yes which products:________________
How many drinks do you drink daily?____/day Years of regular drinking _______ years Do you have / had alcohol addiction? ☐ Yes ☐ No
Drug Use: ☐ Yes ☐ No if yes which products:__________________
Do you have / had drug addiction? ☐ Yes ☐ No
Home /Work
Are your stress levels acceptable to you? ☐ Yes ☐ No
Exposure to toxins, irritants, allergens, etc. in your employment or home? ☐ Yes ☐ No
If "yes", please indicate how and when _________________________

(E) **Physical activity and exercise history**

Hours per week devoted to sedentary activities?________
Present state of physical fitness?
☐ Poor ☐ Below Average ☐ Average ☐ Above Average ☐ Excellent
Do you have a regular exercise program? ☐ Yes ☐ No
If so, what is it? ____
Strenuous sports activities? (Running, swimming, etc.) ☐ Yes ☐ No
If "yes", what activities: ___________________________________

References

1. Vuori I. The cardiovascular risk of physical activity. Acta Med Scan Suppl. 1986;711:205–14.
2. Wang CY, Haskell WL, Farrel SW, et al. Cardiorespiratory fitness levels among US adults 20–49 years of age: finding from the 1999–2004 National Health and Nutrition Examination Survey. Am J Epidemiol. 2010;171(4):426–35.
3. Mittleman MA, Maclure M, Tofler GH, et al. Triggering of acute myocardial infarction by heavy physical exertion. Protection against triggering by regular exertion. Determinants of Myocardial Infarction Onset Study Investigators. N Engl J Med. 1993;329(23):1677–83.
4. Siscovick DS, Weiss NS, Fletcher RH, et al. The incidence of primary cardiac arrest during vigorous exercise. N Engl J Med. 1984;311(14):874–7.
5. Willich SN, Lewis M, Lowel H, et al. Physical exertion as a trigger of acute myocardial infarction. Triggers and Mechanisms of Mycardial Infarction Study Group. N Engl J Med. 1993;329(23):1684–90.
6. Carlsen K-H, Delgado L, Del Giacco S. Diagnosis, prevention and treatment of exercise-related asthma, respiratory and allergic disorders in sports. Eur Respir Monograph. 2005;10. https://doi.org/10.1183/1025448x.erm3305.
7. Global Initiative for Chronic Obstructive Lung Disease. Global strategy for the diagnosis, management, and prevention of chronic obstructive lung disease updated 2018. www.goldcopd.org/guidelines-globalstrategy-for-diagnosis-management.html.
8. Miller RM, Hankinson J, Brusasco V, et al. Standardisation of spirometry. Eur Respir J. 2005;26:319–38.
9. Gibbons RJ, Balady GJ, Bricker JT, et al. ACC/AHA 2002 guideline update for exercise testing: summary article: a report of the American College of Cardiology/American Heart Association Task Force on Practice Guidelines (Committee to Update the 1997 Exercise Testing Guidelines). Circulation. 2002;106(14):1883–92.
10. Thompson PD. Preparticipation in health screening. In: Pescatello LS, Arena R, Riebe D, Thompson PD, editors. ACSM's guidelines for exercise testing and prescription. 9th ed. Philadelphia, PA: Wolters Kluver/Lippincott & Williams and Wilkins; 2014. p. 19–38.
11. Caspersen CJ, Powell KE, Christenson GM. Physical activity, exercise, and physical fitness: Definitions and distinctions for health-related research. Public Health Rep. 1985;100(2):126–31.
12. Bartsch P, Swenson ER. Acute high altitude illnesses. N Engl J Med. 2013;368:2294–302.
13. British Thoracic Society Fitness to Dive Group. British Thoracic Society guidelines on respiratory aspects of fitness for diving. Thorax. 2003;58:3–13.
14. Bove A. Diving medicine. Am J Respir Crit Care Med. 2014;189:1479–86.
15. Lagerros YT, Lagiou P. Assessment of physical activity and energy expenditure in epidemiological research of chronic diseases. Eur J Epidemiol. 2007;22(6):353–62.
16. Valanou EM, Bamia C, Trichopoulou A. Methodology of physical-activity and energy-expenditure assessment: a review. J Public Health. 2006;14(2):58–65.
17. Nelson ME, Rejeski WJ, Blair SN, et al. Physical activity and public health in older adults: recommendation from the American College of Sports Medicine and the American Heart Association. Med Sci Sports Exerc. 2007;39:1435–45.
18. Garcia-Aymerich J, Lange P, Benet M, et al. Regular physical activity reduces hospital admission and mortality in chronic obstructive pulmonary disease: a population based cohort study. Thorax. 2006;61:772–8.
19. American College of Sports Medicine Position Stand. The recommended quantity and quality of exercise for developing and maintaining cardiorespiratory and muscular fitness, and flexibility in healthy adults. Med Sci Sports Exerc. 1998;30:975–91.

20. Leenders NYJM, Sherman M, Nagaraja HN. Comparisons of four methods of estimating physical activity in adult women. Med Sci Sports Exerc. 2000;32:1320–6.
21. Meek PM, Lareau SC, Anderson D. Memory for symptoms in COPD patients: how accurate are their reports? Eur Respir J. 2001;18:474–81.
22. Liesker JJ, Postma DS, Beukema RJ, et al. Cognitive performance in patients with COPD. Respir Med. 2004;98:351–6.
23. Washburn RA, Heath GW, Jackson AW. Reliability and validity issues concerning large-scale surveillance of physical activity. Res Q Exerc Sport. 2000;71(Suppl. 2):S104–13.
24. Follick MJ, Ahern DK, Laser-Wolston N. Evaluation of a daily activity diary for chronic pain patients. Pain. 1984;19:373–82.
25. Friedenreich CM, Courneya KS, Bryant HE. The lifetime total physical activity questionnaire: development and reliability. Med Sci Sports Exerc. 1998;30:266–74.
26. Bonnefoy M, Normand S, Pachiaudi C, et al. Simultaneous validation of ten physical activity questionnaires in older men: a doubly labeled water study. J Am Geriatr Soc. 2001;49:28–35.
27. Jacobs DR Jr, Ainsworth BE, Hartman TJ, Leon AS. A simultaneous evaluation of 10 commonly used physical activity questionnaires. Med Sci Sports Exerc. 1993;25:81–91.
28. Sarkin JA, Nichols JF, Sallis JF, et al. Self-report measures and scoring protocols affect prevalence estimates of meeting physical activity guidelines. Med Sci Sports Exerc. 2000;32:149–56.
29. Schutz Y, Weinsier RL, Hunter GR. Assessment of free living physical activity in humans: an overview of currently available and proposed new measures. Obes Res. 2001;9:368–79.
30. Vanhees L, Lefevre J, Philippaerts R, et al. How to assess physical activity? Eur J Cardiovasc Prev Rehabil. 2005;12:102–14.
31. Pitta F, Troosters T, Probst VS, et al. Quantifying physical activity in daily life with questionnaires and motion sensors in COPD. Eur Respir J. 2006;27:1040–55.
32. Last JM. A dictionary of epidemiology. 3rd ed. New York: Oxford University Press Inc.; 1995.
33. Kriska A. Ethnic and cultural issues in assessing physical activity. Res Q Exerc Sport. 2000;71(Suppl 2):S47–53.
34. Uitenbroek DG. Seasonal variation in leisure time physical activity. Med Sci Sports Exerc. 1993;25:755–60.
35. Washburn RA, Smith KW, Jette AM, et al. The physical activity scale for the elderly (PASE): development and evaluation. J Clin Epidemiol. 1993;46:153–62.
36. Folsom AR, Jacobs DR Jr, Caspersen CJ, et al. Test–retest reliability of the Minnesota leisure time physical activity questionnaire. J Chronic Dis. 1986;39:505–11.
37. Caspersen CJ, Bloemberg BP, Saris WH, et al. The prevalence of selected physical activities and their relation with coronary heart disease risk factors in elderly men: the Zutphen study, 1985. Am J Epidemiol. 1991;133:1078–92.
38. Voorrips LE, Ravelli AC, Dongelmans PC, et al. A physical activity questionnaire for the elderly. Med Sci Sports Exerc. 1991;23:974–9.
39. Trost SG, Pate RR, Freedson PS, et al. Using objective physical activity measures with youth: how many days of monitoring are needed? Med Sci Sports Exerc. 2000;32:426–31.
40. Guillemin F, Bombardier C, Beaton D. Cross-cultural adaptation of health-related quality of life measures: literature review and proposed guidelines. J Clin Epidemiol. 1993;46:1417–32.
41. Guyatt GH, Kirshner B, Jaeschke R. Measuring health status: what are the necessary measurement properties? J Clin Epidemiol. 1992;45:1341–5.
42. Garrod R, Paul EA, Wedzicha JA. An evaluation of the reliability and sensitivity of the London chest activity of daily living scale (LCADL). Respir Med. 2002;96:725–30.
43. Yohannes AM, Roomi J, Winn S, et al. The Manchester respiratory activities of daily living questionnaire: development, reliability, validity, and responsiveness to pulmonary rehabilitation. J Am Geriatr Soc. 2000;48:1496–500.

44. Bendstrup KE, Ingemann JJ, Holm S, et al. Outpatient rehabilitation improves activities of daily living, quality of life and exercise tolerance in chronic obstructive pulmonary disease. Eur Respir J. 1997;10:2801–906.
45. Williams K, Frei A, Vetsch A, et al. Patient-reported physical activity questionnaires: a systematic review of content and format. Health Qual Life Outcomes. 2012;10:28–45.
46. Frei A, Williams K, Vetsch A, et al. A comprehensive systematic review of the development process of 104 patient-reported outcomes (PROs) for physical activity in chronically ill and elderly people. Health Qual Life Outcomes. 2011;9:116.
47. Forsen LLN, Vuillemin A, Chinapaw MJM, et al. Self-administered physical activity questionnaires for the elderly: a systematic review of measurement properties. Sports Med. 2010;40:601–23.
48. Gimeno-Santos E, Raste Y, Demeyer H, et al. The PROactive instruments to measure physical activity in patients with chronic obstructive pulmonary disease. Eur Respir J. 2015;46:988–1000.
49. Dobbels F, de Jong C, Drost E, et al. The PROactive innovative conceptual framework on physical activity. Eur Respir J. 2014;44:1223–33.
50. Pinnock H, Burton C, Campbell S, et al. Clinical implications of the Royal College of Physicians three questions in routine asthma care: a real-life validation study. Prim Care Respir J. 2012;21:288–94.
51. Juniper EF, O'Byrne PM, Guyatt GH, et al. Development and validation of a questionnaire to measure asthma control. Eur Resp J. 1999;14:902–7.
52. Juniper EF, Svensson K, Mork AC, et al. Measurement properties and interpretation of three shortened versions of the asthma control questionnaire. Respir Med. 2005;99:553–8.
53. Fletcher CM. Standardised questionnaire on respiratory symptoms: a statement prepared and approved by the MRC Committee on the Aetiology of Chronic Bronchitis (MRC breathlessness score). BMJ. 1960;2:1662.
54. Bestall JC, Paul EA, Garrod R, et al. Usefulness of the Medical Research Council (MRC) dyspnoea scale as a measure of disability in patients with chronic obstructive pulmonary disease. Thorax. 1999;54(7):581–6.
55. Sundh J, Janson C, Lisspers K, et al. The dyspnoea, obstruction, smoking, exacerbation (DOSE) index is predictive of mortality in COPD. Prim Care Respir J. 2012;21(3):295–301.
56. Nishimura K, Izumi T, Tsukino M, et al. Dyspnea is a better predictor of survival than airway obstruction in patients with COPD. Chest. 2002;121(5):1434–40.
57. Jones PW, Quirk FH, Baveystock CM, et al. A self-complete measure of health status for chronic airflow limitation. The St. George's respiratory questionnaire. Am Rev Respir Dis. 1992;145(6):1321–7.
58. Jones PW, Harding G, Berry P, et al. Development and first validation of the COPD assessment test. Eur Respir J. 2009;34(3):648–54.
59. Karloh M, Fleig Mayer A, Maurici R, et al. The COPD assessment test: what do we know so far?: a systematic review and meta-analysis about clinical outcomes prediction and classification of patients into GOLD stages. Chest. 2016;149(2):413–25.
60. Bonini M, Braido F, Baiardini I, et al. AQUA: allergy questionnaire for athletes. development and validation. Med Sci Sports Exerc. 2009;41:1034–41.
61. Bickley LS, Szilagyi PG. Bates' guide to physical examination and history taking. 11th ed. Philadelphia, PA: Wolters Kluwer Health/Lippincott, Williams & Wilkins; 2013.
62. Talley NJ, O'Connor S. Clinical examination a systematic guide to physical diagnosis. 7th ed. NSW: Elsevier; 2013.
63. Boulet L-P, O'Byrne PM. Asthma and exercise-induced bronchoconstriction in athletes. N Engl J Med. 2015;372:641–8.
64. Parsons JP, Hallstrand TS, Mastronarde JG, et al. An Official American Thoracic Society Clinical Practice Guideline: exercise-induced bronchoconstriction. Am J Respir Crit Care Med. 2013;187:1016–27.

65. Del Giacco SR. Exercise-induced anaphylaxis: an update. Breathe. 2012;8:298–306.
66. Brignole M, Moya A, de Lange FJ, et al. ESC Guidelines for the diagnosis and management of syncope. The task force for the diagnosis and management of syncope of the European Society of Cardiology (ESC). Eur Heart J. 2018;00:1–69.
67. Shen WK, Sheldon RS, Benditt DG, et al. ACC/AHA/HRS Guideline for the evaluation and management of patients with syncope. A Report of the American College of Cardiology/American Heart Association Task Force on Clinical Practice Guidelines and the Heart Rhythm Society. J Am Coll Cardiol. 2017;70(5):e39–e110.

Tests of Lung Function: Physiological Principles and Interpretation

3

Vito Brusasco and Giovanni Barisione

Abstract

Gas exchange between organism and external ambient is the ultimate task of the respiratory system. Its efficiency is critically dependent on the efficiencies of ventilation and gas transport across the airspaces and lung tissues. Therefore, the knowledge of physiological principles underlying tests of lung function at different levels is basic to the understanding of the mechanisms limiting respiratory efficiency under different conditions, such as exercise and disease. The first step of lung function testing in clinical practice is spirometry, but it does not allow distinguishing the causes of airflow obstruction, i.e. airway disease versus emphysema, or establishing a diagnosis of lung restriction. Moreover, the effects of volume history and thoracic gas compressions may complicate its interpretation. Therefore, measurements of lung volumes are often necessary not only to confirm restriction in subjects with restrictive spirometric pattern but also for the assessment of lung hyperinflation. The latter may be due to either static (loss of elastic recoil) or dynamic (airflow limitation) mechanisms. The inhomogeneity of lung mechanics can be assessed by forced oscillations and/or nitrogen washout and may be more sensitive than spirometry to early obstruction of peripheral airways. The final step of lung function testing in clinical practice is the assessment of lung diffusing capacity for carbon monoxide. This test reflects the transport of gases from airspaces to blood across the alveolar-to-capillary barrier. Its interpretation is not always easy because the major resistance to carbon monoxide transfer is in the red cells rather than in the alveolar-to-capillary membrane.

V. Brusasco (✉) · G. Barisione
U.O. Fisiopatologia Respiratoria, Dipartimento di Medicina Interna, Università di Genova, Ospedale Policlinico San Martino, Genoa, Italy
e-mail: vito.brusasco@unige.it; giovanni.barisione@hsanmartino.it

A. Cogo et al. (eds.), *Exercise and Sports Pulmonology*,
https://doi.org/10.1007/978-3-030-05258-4_3

3.1 Introduction

In any animal species, life requires energy consumption, which is provided by oxygen (O_2) uptake from the external ambient, and carbon dioxide (CO_2) elimination. In mammals, the first step of this process takes place in the lung. During physical activity, metabolic requirements are manifold greater than at rest, thus requiring the respiratory system to increase rapidly its performance. Ventilation is the process of moving the air in and out of the lungs so that gas exchange can occur between the environmental air and the body. The efficiency of ventilation depends on the activity of respiratory muscles and the passive mechanical properties of the respiratory system, mainly stemming from its elastic and resistive structures. Gas exchange is the primary function of the lung and must be able to adapt to different life conditions, such as hypoxia or heavy exercise. The efficiency of gas exchange depends on the efficiency of ventilation, the surface and thickness of air-blood barrier, the magnitude of pulmonary blood flow and haemoglobin concentration ([Hb]).

This chapter will summarize the physiological principles underlying measurements of respiratory mechanics and gas exchange to provide practical keys for their interpretation in health and disease, at rest and on exercise.

3.2 Static Properties of Respiratory System

3.2.1 Subdivisions and Determinants of Lung Volumes

Total lung capacity (TLC) is the volume of air in the lungs at the end of a maximal inspiration and is determined by the static balance between the decreasing force of inspiratory muscles and the increasing inward *recoil of the respiratory system* ($P_{el,rs}$) [1]. In normal subjects at full lung inflation, the inward *recoil pressure of the lung* ($P_{el,L}$) largely exceeds the *recoil pressure of the chest wall* ($P_{el,cw}$), and the pressure–volume (P–V) curve of the lung but not the chest wall exhibits a plateau. Thus, the limit of TLC is represented by the passive characteristics of lung tissue, although elite breath-hold divers can inhale substantial amount of air beyond their normal TLC by *genioglossal insufflation* [2], a phenomenon in part accounted for by gas compression [3] and ability of their lungs to withstand transpulmonary pressure (P_{tp}) values much larger than normal. TLC does not vary substantially with ageing, likely because the decrease in $P_{el,L}$ is balanced by an increase in chest wall stiffness and/or a decrease of inspiratory muscle force [4], and does not change on exercise [5].

An increased TLC can be present in *chronic obstructive pulmonary disease* (COPD), when $P_{el,L}$ is reduced due to emphysema, but it may also occur during severe asthma attacks [6], though the mechanism for this is not clear. A decrease of TLC is the gold standard for the diagnosis of *restrictive disorders*, which can be due to an increase of $P_{el,L}$, or volume shrinkage, or inspiratory muscle weakness [7]. These mechanisms to be distinguished would need measurement of $P_{el,L}$ and maximal inspiratory pressure.

Residual volume (RV) is the volume of air remaining in the lungs after a maximal expiration. In young healthy subjects, its determinant is the static balance between the decreasing force of expiratory muscles and the increasing outward $P_{el,cw}$ [8]. In older subjects and obstructive disorders, this static balance cannot be achieved because airway closure or extreme flow limitation occurs during expiration. The RV/TLC ratio was originally proposed as an index of emphysema, but it may also increase in non-emphysematous air trapping or even in restrictive disorders if RV is proportionally less reduced than TLC.

Vital capacity (VC) is the maximum volume of air that one can mobilize with a single manoeuvre; thus it is the difference between TLC and RV. In young healthy subjects, it is statically determined but in older subjects and obstructive disorders may be limited by the same dynamic mechanism as RV. VC was the first lung volume used clinically [9], and its reduction has been considered as a sign of respiratory disease with prognostic value. A decrease of VC may result from a decrease in TLC or an increase in RV, thus not allowing differentiate between restrictive and obstructive abnormalities [10].

Functional residual capacity (FRC) is the volume of gas remaining in the lungs at the end of a relaxed expiration. In healthy subjects breathing quietly, it is determined by the static balance between the inward $P_{el,L}$ and the outward $P_{el,cw}$ [1], thus corresponding to the *relaxation volume* (V_r) of the respiratory system. In healthy subjects, V_r depends on body position and external pressure acting on chest wall or abdomen. In restrictive disorders, V_r may decrease because of increased $P_{el,L}$ (pulmonary fibrosis, atelectasis, lung resection, alveolar oedema, cardiac diseases) or decreased outward $P_{el,cw}$ (pleural diseases, scoliosis, neuromuscular disorders, obesity) [11]. In obstructive disorders, V_r may increase because of reduced $P_{el,L}$ (emphysema), which defines *static lung hyperinflation*. In normal subjects, FRC is $<V_r$ when expiration is not relaxed but involves expiratory muscle activity, such as during exercise. In obstructive disorders, FRC may be $>V_r$ if expiratory time is reduced [12], or emptying time constant (resistance × compliance) is increased, or expiratory flow is limited during tidal expiration [13, 14]. FRC > V_r defines *dynamic lung hyperinflation*, which is associated with intrinsic positive end-expiratory pressure (PEEPi) [15] and is a major cause of dyspnoea either at rest or on exercise.

Inspiratory capacity (IC) is the maximum volume that can be inspired from the end of a tidal expiration and thus is the difference between FRC and TLC. It has been used as an indirect measure of changes in FRC under conditions where TLC can be assumed to remain constant, e.g. exercise [5] or acute pharmacological interventions [14].

3.2.2 Measurements of Lung Volumes

In vivo, FRC, RV and TLC are measurable by indirect methods only. The classic ones are *multiple-breath nitrogen washout* (MBN_2W), *multiple- or single-breath helium (He) dilution* and *body plethysmography*. Washout and dilution methods are based on the principle of the conservation of mass and measure only the amount of gas that is present in lung regions communicating with open airways [16], provided corrections for N_2 eliminated from body tissues or He dissolved in blood during the tests are applied. Body plethysmography is based on Boyle's law and measures total thoracic gas volume, including regions possibly not communicating with open airways [17]. Radiographic techniques, either standard chest X-ray [18] or computed tomography (CT) [19], provide measurements of lung volumes close to body plethysmography. In healthy subjects and in restrictive abnormalities, all methods give similar results. In severely obstructed subjects, owing to some regions with time constants indistinguishable from those of tissues, the dilution methods may underestimate FRC and TLC [20, 21], whereas plethysmography tends to overestimate them because changes in mouth and airway pressures may differ substantially during panting [22].

3.3 Dynamic Properties of the Respiratory System

3.3.1 Tidal Breathing

Tidal volume (V_T) is the volume of air inspired and expired per breath. In healthy subjects, the size of V_T is regulated by switching off inspiration before end-tidal inspiratory volume achieves the flat part of P–V curve and terminating expiration when V_r is passively achieved [23]. At rest, normal tidal expiration is slightly longer than inspiration, due to braking effects of glottis narrowing and post-inspiratory activity of inspiratory muscles [23]. In obstructive disorders, the difference between inspiratory and expiratory times increases because of the slow emptying time constant [7]. During exercise, V_T increases mostly by increasing end-inspiratory lung volume but also by decreasing FRC, thus maintaining the operative volume over the linear part of P–V curve and limiting the increased work of breathing [5]. The interest of measuring V_T is restricted to exercise and in ventilated patients.

Work of breathing (WOB) is the product of changes in driving pressure and V_T, which is the energy expenditure to move air in and out of the lungs. In healthy subjects breathing quietly, inspiration requires work by inspiratory muscles to overcome the $P_{el,rs}$ and, to lesser extent, the resistance to airflow $(\dot{V})$, whereas expiration is passive, thus requiring no work by expiratory muscles [23]. During exercise, the elastic WOB increases proportionally to V_T and the resistive WOB proportionally to $\dot{V}$ [5]. In restrictive disorders, the elastic WOB increases on inspiration because of the increase of $P_{el,rs}$. In obstructive disorders with dynamic lung hyperinflation, the elastic WOB also increases because the respiratory system operates in the upper part of the P–V curve, where $P_{el,L}$ is large, and additional pressure may be required to overcome PEEPi. In severe airflow obstruction, also resistive WOB may increase, particularly when expiration is completed by the activity of expiratory muscles. Thus, irrespective of the underlying mechanisms, the WOB may increase in both restrictive and obstructive disorders [23].

Respiratory system resistance (R_{rs}) is the sum of all pressures dissipated across airway, lung parenchyma and chest wall divided by $\dot{V}$. R_{rs} can be measured by superimposing high-frequency forced oscillations to tidal breathing [24]. In healthy subjects, R_{rs} is independent of oscillation frequency, whereas in subjects with obstructive disorders is increased at low frequencies and decreases at high frequencies. Modelling studies have suggested that such a frequency dependence is due to central airway resistance dominating at high frequencies with inhomogeneity and peripheral resistance dominating at low frequencies. This method also provides the reactance of the respiratory system (X_{rs}), which reflects gas compliance at low frequency and gas inertance at high frequencies. X_{rs} is negative at low frequencies and more so in the presence of reduced lung volume (e.g. restrictive disorders) or ventilation inhomogeneity (e.g. obstructive disorders). Moreover, a more negative X_{rs} on expiration than inspiration is suggestive of EFL [25, 26].

Airway resistance (R_{aw}) is the ratio of alveolar-to-mouth pressure difference and $\dot{V}$. During tidal breathing, most of R_{aw} is in the upper airways [27] with large intra- and interindividual variability, likely due to anatomical differences. The contribution of larynx to R_{aw} is greater on expiration than inspiration, because its calibre increases during the latter. R_{aw} can be measured by body plethysmography [17] or interruptor technique [28]. With the former, changes of alveolar pressure (PA) are assumed to be equal to changes in box volume or pressure, and laryngeal resistance can be minimized by panting, during which vocal cords are maximally abducted. With the latter, PA is assumed to be equal to the mouth pressure drop following a rapid airway occlusion, which may not be fully true because of damping in lung and chest wall tissues. In healthy subjects, R_{aw} is virtually independent of breathing frequency but highly dependent on lung volume, because the calibre of intraparenchymal airways varies approximately with the cube root of volume and the resistance to laminar $\dot{V}$ is inversely related to the fourth power of airway radius [29]. The inverse of R_{aw}, i.e. airway conductance (G_{aw}), increases linearly with lung volume; thus specific airway conductance ($sG_{aw} = sG_{aw}/FRC$) is a reasonable correction for lung volume.

Pulmonary resistance (R_L) is the ratio of P_{tp} to $\dot{V}$ and includes R_{aw} and tissue resistance (R_{ti}). P_{tp} can be measured by an oesophageal balloon and R_L derived from the classic equation of motion [23]. R_L decreases with breathing frequency, due to the frequency dependence of R_{ti}, and lung volume, due to the volume dependence of R_{aw} [29]. The contribution of peripheral R_{aw} to total R_L is 25–30% in healthy subjects and 50–60% in COPD [30]. The use of R_L for clinical purposes is limited.

3.3.2 Forced Manoeuvres

Forced expiration was introduced in 1947 by Tiffeneau and Pinelli [31] and has become the mainstay of pulmonary function testing. Studies on the relationships between flow, volume and pressure showed that most of forced expiratory manoeuvre is effort independent [32], thus confirming its validity as a test of lung function.

Expiratory flow limitation (EFL) is a phenomenon occurring during forced expiration in all mammals' lungs, either healthy or diseased. Its evidence stems from the observation that, at lung volumes <75% of VC, forced expiratory $\dot{V}$ increases with pressure up to a point above which it cannot further increase with increasing effort [32, 33]. EFL, initially explained with waterfall analogy [34] and equal-pressure point theory [35], was subsequently enlightened by the *wave-speed theory* [36]. Briefly, because airways are collapsible tubes, the maximum expiratory $\dot{V}$ $\left(\dot{V}_{\mathrm{E}}\,\mathrm{max}\right)$, at a given point of each airway, is determined by cross-sectional area (A), wall elastance ($E = \Delta P/\Delta A$) and gas density (ρ):

$$\dot{V}_{\mathrm{E}}\,\mathrm{max}/\,A = \left(\frac{A}{\rho}\cdot\frac{\Delta P}{\Delta A}\right)^{1/2}$$

The point at which this phenomenon occurs along the airways is called *choke point.* Because P is a function of lung volume and P_{L}, the choke point moves from the trachea at high lung volume down the bronchial tree as the lung empties during forced expiration and progressively to the more compliant peripheral airways. For this reason, $\dot{V}_{\mathrm{E}}\,\mathrm{max}$ at mid-to-low lung volumes was proposed as more sensitive than the usual spirometric indices for early detection of *small airway disease*. However, their clinical usefulness is not demonstrated [37]. Moreover, $\dot{V}_{E}\,\mathrm{max}$ can be equally reduced whether P is decreased due to frictional losses upstream from the limiting segment (small-airway obstruction) or loss of $P_{\mathrm{el,L}}$ (emphysema), which makes it impossible to separate the mechanisms of EFL by simple spirometry. Because of the inverse relationship between $\dot{V}_{\mathrm{E}}\,\mathrm{max}$ and ρ, breathing a low-density mixture, e.g. 80% He and 20% O_2, can increase $\dot{V}_{\mathrm{E}}\,\mathrm{max}$ at high-to-mid lung volumes, when the flow-limiting segment is in the large airways where P drops due to convective acceleration and turbulent flow. A reduction of density dependence was thus proposed as a sign of obstruction of peripheral airways, where $\dot{V}$ is laminar and viscosity-dependent [38]. However, unaltered density dependence was observed in severe COPD [39], possibly because it reflects more the ratio of lung volume to central airway calibre rather than the latter by itself [40].

Effects of volume history and thoracic gas compression occur during forced expiration manoeuvre and may profoundly influence spirometric measurements (Fig. 3.1). *First*, full lung inflation to TLC may cause transient changes of airway calibre, namely, *bronchodilation during induced bronchoconstriction* [41] and *bronchoconstriction during spontaneous asthma* [42] or *COPD* [43]. *Second*, because of EFL and gas compression, changes in expired volume lag true changes in lung volume. This effect, which can be shown by plotting $\dot{V}_{\mathrm{E}}\,\mathrm{max}$ measured at the mouth against expired or plethysmographic volume, is small in healthy subjects but may be large when R_{aw} and lung volume are increased [44] and may amplify changes in airway calibre [45]. Because of these effects, the forced expiratory

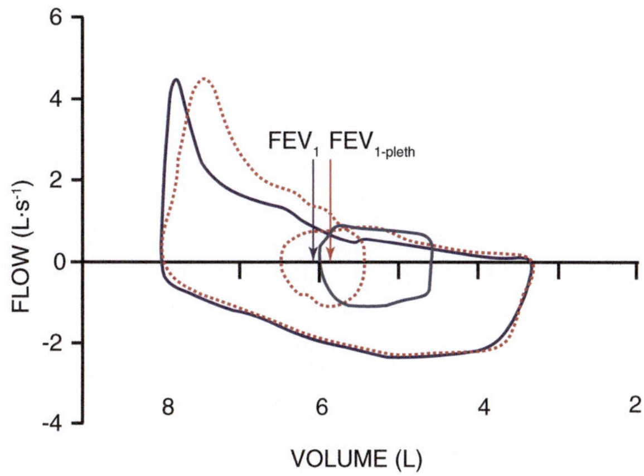

Fig. 3.1 Representative tidal and maximal flow-volume curves in emphysema. Flow is plotted against expired volume (*continuous line*) or thoracic volume measured by plethysmography (*dotted line*). Forced expiratory volume in 1 s measured by plethysmography ($FEV_{1\text{-pleth}}$) is larger than at the mouth (FEV_1). Note that tidal expiratory flow exceeds forced expiratory flow when plotted against expired but not against thoracic volume

volume in 1 s (FEV_1) may either overestimate or underestimate the degree of airway narrowing and drug-induced changes.

EFL and tidal breathing. Normally, tidal breathing requires a $\dot{V}$ much less than $\dot{V}\max$, even on exercise. Only in extremely fit athletes and in subjects affected by respiratory diseases, tidal $\dot{V}_E$ may impinge on $\dot{V}_E\max$, suggesting breathing under EFL conditions, though this finding may be in part due to volume history and gas compression. Thus, the gold standard for identifying EFL during tidal breathing is by showing no increase in tidal $\dot{V}_E$ with increasing intrapleural pressure (P_{pl}). Alternative non-invasive methods proposed for identification of EFL during tidal breathing void of volume history and gas compression effects include the comparison of tidal and submaximal $\dot{V}_{\text{E}}$ obtained voluntarily [46] or by manual compression of the abdominal wall [47], application of expiratory negative pressure at mouth [48], interrupt or technique [49] and reduction of X_{rs} during expiration [25, 26]. For practical purposes, owing to the association between EFL and dynamic lung hyperinflation [46], changes in FRC or IC following bronchodynamic interventions (e.g. bronchial challenge or reversibility test) or during exercise may be taken as suggestive of tidal EFL.

Forced inspiration is mostly dependent on the force of inspiratory muscles and their shortening velocity. At 50% of FVC, inspiratory $\dot{V}\max$ slightly exceeds $\dot{V}_{\text{E}}\max$ in normal subjects; it is reduced more than $\dot{V}_{\text{E}}\max$ in case of variable extrathoracic obstruction and similarly decreased in fixed obstruction [50].

3.4 Distribution of Ventilation and Pulmonary Gas Exchange

In healthy lungs, each tidal inspiration (~500 mL) at rest brings a column of fresh air by convection into the acinus and by convection plus diffusion up to the entrance of alveolar ducts [51]. Airflow velocity falls from ~1 m· s^{-1} in the trachea to <1 cm· s^{-1} in the first order respiratory bronchioles because the airway total cross-sectional area increases with every generation. In exercise, flow velocities are up to ten times greater, in proportion to the increased ventilation ($\dot{V}$).

At FRC, there is a gradient of lung expansion from gravity-dependent to nondependent lung regions. On expiration to RV, the former reach their minimum volume before the latter. This phenomenon is due to regional gravity-dependent *small-airway closure* and is more prominent in the elderly [52] and in the early phase of COPD due to the decline of $P_{el,L}$ [53]. Non-uniform alveolar $\dot{V}$ ($\dot{V}_A$) has physiological and clinical significance, both intrinsically (gas mixing inefficiency) and because the distribution of $\dot{V}_A$ is a determinant of ventilation-to-perfusion ratio ($\dot{V}_A / \dot{Q}$).

3.4.1 Measurements of Ventilation Heterogeneity

Single-breath N_2 washout (SBN_2W) is a simple test requiring an inspiration of 100% O_2 from RV to TLC followed by a slow expiration to RV [53, 54]. The resulting percentage change in N_2 concentration versus expired volume curve ([ΔN_2]%·L^{-1}) has four phases with *phase I* representing the N_2-free anatomic dead space, *phase II* the fast rising [N_2] from the transition zone between airways and alveolar space, *phase III* the slowly rising [N_2] from alveolar space and *phase IV* a sharp increase of [N_2] from apical zones. The slope of phase III (S_{III}) is in large part gravity-independent [55] and considered to reflect uneven emptying of adjacent lung regions, possibly at small-airway level [56]. By contrast, phase IV represents the closing volume (CV) of gravity-dependent lung regions. The CV is of physiological interest but not particularly useful for clinical practice [57], and the transition from phase III to IV is often not detectable in obstructed patients.

Multiple-breath N_2 washout (MBN_2W) was first introduced in 1940 to measure FRC [58]. However, more information than just FRC can be derived from the analysis of N_2 exponential decay curve. In particular, the volume of pure O_2 that must be breathed in order to lower expired [N_2] to 1/40th of initial value is a simple parameter describing gas mixing efficiency [59]. Correcting for lung size using FRC allows derivation of the *lung clearance index* (LCI) [60], which, despite its wide biological variability [61], has been shown to be more sensitive than spirometry or airflow resistance measurements in detecting lung function abnormalities in children with cystic fibrosis [62]. A major advance in the field was the introduction of the analysis of S_{III} of the first 20 breaths of MBN_2W [63]. This approach has the advantages over the SBN_2W of distinguishing convective-dependent inhomogeneity at airway branching points proximal to the entrance to the acinus from diffusive-convective-dependent inhomogeneity located within or between acini [64].

3.4.2 Alveolar-to-Capillary Gas Exchange

The sandwich of tissues separating the alveolar spaces from blood in the lung must be crossed bidirectionally by O_2 and CO_2 passive molecular diffusion. The diffusive resistance offered by alveolar-capillary membrane is very small, largely because this is exceedingly thin, i.e. a fraction of μm on average, and extremely large, i.e. ~50–150 m^2 [65]. However, the diffusive resistance to O_2 uptake may become measurable under various conditions leading to *diffusion limitation* of alveolar gas exchange and arterial hypoxemia. In clinical practice, carbon monoxide (CO) instead of O_2 is used for measurements of lung diffusing capacity (DL_{CO}) owing to its very high capacitance coefficient (β) and low (<3) $D/(\dot{Q}\beta)$ ratio [66]. A value of alveolar-capillary O_2 transfer can be, therefore, calculated based on a DL_{O_2}/DL_{CO} ratio of ~1.61 [67]. Basically, DL_{CO} measures surface area and thickness of the air-blood barrier available for gas exchange. In health, it represents the *upper bound* value for alveolar-capillary gas exchange. CO entering pulmonary capillary blood binds to the haem group of circulating Hb as COHb at an extremely low and unmeasurable capillary partial pressure. Hence, CO uptake is independent of pulmonary $\dot{Q}$ but critically dependent on capillary blood volume and [Hb] [68].

Single-breath DL_{CO} is used in clinical laboratories, with a breath-hold time of 9–11 s at maximal inspiration (~TLC). This technique was originally described by Marie Krogh in 1915 [69] and subsequently modified with the addition of He as marker inert gas [70]. Its technical standards [71] and reference values [72] have been recently updated. In brief, both CO and the relevant marker gas (He or CH_4) mix immediately after inhalation with the gas resident in the lung at ~RV. During the subsequent breath-hold at TLC, CO is taken up from the alveoli, but He and CH_4 are not. Therefore, the [He] or [CH_4] in alveolar gas on expiration can be back-extrapolated to the effective time zero and to the initial alveolar [CO] before uptake has occurred. DL_{CO} is, then, obtained as follows:

$$DL_{CO} = k_{CO} \cdot V_A$$

where k_{CO} is the rate constant of CO alveolar uptake during the breath-hold and V_A the alveolar volume, i.e. the maximal lung volume available for dilution to inhaled CO and He or CH_4 [73]. Because the measurement is made at full lung inflation, V_A is in the presence of low gas phase resistance within 5–10% of TLC measured by plethysmography. K_{CO} is the transfer factor per unit alveolar volume and mathematically equal to DL_{CO}/V_A but is not DL_{CO} corrected for lung volume because K_{CO} is actually the rate constant k_{CO} normalized to barometric pressure. Thus, K_{CO} is better regarded as the transfer factor per alveolus or per acinus, representing the efficiency of the CO alveolar uptake [73]. Single-breath DL_{CO} is measured during an unphysiological manoeuvre (breath-holding at ~TLC), but this has the advantages of *(a)* a "reproducible" lung volume and *(b)* optimization of the test gas distribution. However, the latter condition is hardly ever achieved in the presence of airflow obstruction where the distribution of test gases to peripheral gas-exchange units is variably compromised. Thus, in COPD patients V_A is often severely underestimated, and DL_{CO} may be considered the *lower bound* value of effective pulmonary gas exchange [73].

3.4.3 Subcomponents of Lung Diffusing Capacity

Anatomically, the blood-gas barrier encompasses the surfactant lining layer, alveolar epithelium, interstitium, capillary endothelium, plasma, erythrocyte plasmalemma and the Hb molecule within erythrocytes [65].

Two-step PA_{O_2} *method.* In 1957, Roughton and Forster [74] found that, owing to competitive binding between CO and O_2 for Hb-accessible sites, DL_{CO} fell systematically when O_2 alveolar partial pressure (PA_{O_2}) was increased. According to their method, the DL_{CO} measured at PA_{O_2} of 100 and 500 mmHg, respectively, may be graphically partitioned into two resistances arranged in series as follows:

$$\frac{1}{DL_{CO}} = \frac{1}{DM_{CO}} + \frac{1}{\vartheta_{CO} \cdot Vc}$$

where DM_{CO} represents the membrane diffusive conductance for CO, ϑ_{CO} is the rate of reaction of CO with deoxy-Hb in the red blood cell and Vc is the pulmonary capillary blood volume. The latter two subcomponents constitute the *erythrocyte conductance* (De_{CO}) which is the *reactive* or O_2-dependent part of the transfer resistance. Following this approach, both $DM_{CO}$and De_{CO} are estimated to be ~50% each at rest and ~80% and ~122%, respectively, on exercise [75]. However, nonlinearity of the $1/\vartheta_{CO}$- PA_{O_2} relationship could lead to an overestimation of the zero PA_{O_2} *y*-intercept and underestimation of DM_{CO} suggesting that 75–80% of the transfer resistance is due to $1/De_{CO}$, which is in accordance with calculations based on morphometric data [65].

Simultaneous uptake of nitric oxide (NO) and CO. This method takes advantage of the extremely rapid and greater than CO chemical combination of NO with Hb [76] to solve the Roughton-Forster equation with a single manoeuvre [77, 78]. DL_{NO} approximates the morphometric value of DL_{O_2} [65], whereas the DL_{NO}/DL_{CO} ratio is ranging from 4.3 to 5.3 [77, 78]. Although it was demonstrated that there is a *significant* $1/De_{NO}$ [79], peri- and intraerythrocyte resistance to NO uptake is negligible as compared to CO, and DL_{NO} can be regarded operationally as a surrogate for DM_{NO} [80].

3.5 Interpretative Strategies

Clinical practice recommendations jointly published by the American Thoracic Society and the European Respiratory Society provide a flowchart to assess spirometry, lung volumes and DL_{CO} results [81]. The interpretation of pulmonary function

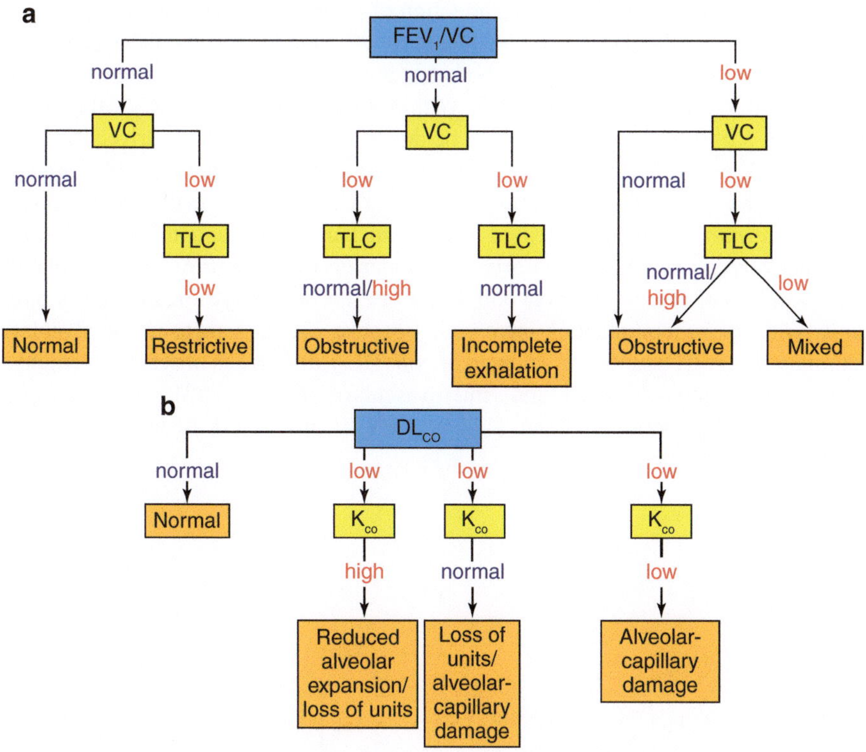

Fig. 3.2 Simplified interpretative algorithms for spirometry and lung volumes (Panel a), and lung diffusing capacity (Panel b) in clinical practice. FEV_1, forced expiratory volume in 1 s; VC, vital capacity, TLC, total lung capacity, DL_{CO}, lung diffusing capacity for carbon monoxide adjusted for effective [Hb], K_{CO}, carbon monoxide transfer factor per unit alveolar volume (DL_{CO}/V_A)

tests is based on the comparison of measurements with the reference values of healthy people of the same sex, age and height.

Spirometry and lung volumes are the first step in the interpretation of pulmonary function tests (Fig. 3.2, panel a), starting from the *FEV_1/VC ratio*. If this is normal and VC is below the fifth percentile of the normal distribution (LLN), then *restriction* may be suspected, but it requires to be confirmed by a reduction of TLC. Otherwise, the reduction of VC may be the result of an increase in RV, due to either *air trapping* (early obstruction) or *reduced force of expiratory muscles*. If FEV_1/VC is <LLN, then *airflow obstruction* is present but may be associated with either normal or reduced VC. When the latter is the case, measurement of TLC is required to say whether the reduced VC is due to air trapping or a *mixed obstructive-restrictive* abnormality. A reversibility test with an inhaled bronchodilator may help distinguish between *fully reversible* (generally present in asthma) and *fixed* (generally present in COPD) airflow obstruction. However, a partial response does not allow to differentiate between asthma and COPD or predict the efficacy of long-term bronchodilator treatment [81].

DL_{CO} is the final step in the interpretation of pulmonary function tests (Fig. 3.2, panel b). Because the DL_{CO} is the product of the K_{CO} and V_A, its final value can result from a number of combinations of both K_{CO} and V_A, each pattern being associated with a different pathological process. With *normal or increased* DL_{CO}, the *increment of* K_{CO} observed on exercise may be due to the pulmonary $\dot{Q}$-driven increase of Vc and Vc/V_A [82]. On the other hand, a *reduced* DL_{CO} with a K_{CO} *substantially increased* may be seen with a low (e.g. <0.85) V_A/TLC ratio and results from *reduced alveolar expansion* with decreased DM_{CO}, unchanged Vc and increased Vc/V_A. Alternatively, a *low* DL_{CO} *with normal or moderately increased* K_{CO} may be due to localized loss of alveolar units in both lungs or after lung resection, i.e. *loss of lung units* with all subcomponents increased. Because K_{CO} is a measure of alveolar-capillary integrity, a mutual *decrement of both* DL_{CO} *and* K_{CO} is observed in emphysema, pulmonary fibrosis and microvascular damage [73]. The relevant interrelationships between V_A and K_{CO} and with the different diffusion components and subcomponents are shown in Fig. 3.3.

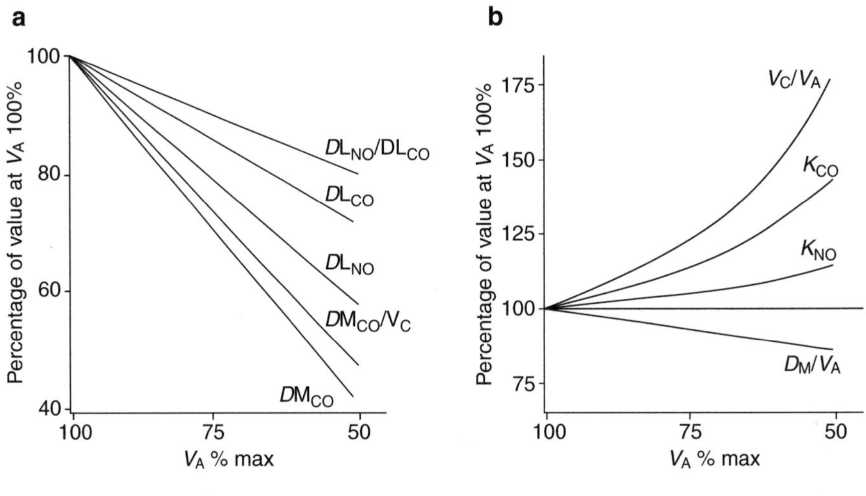

Fig. 3.3 Plots of (**a**) lung diffusing capacity for nitric oxide (DL_{NO}) and carbon monoxide (DL_{CO}), their ratio (DL_{NO}/DL_{CO}), alveolar-capillary membrane diffusing capacity for CO (DM_{CO}) and the DM_{CO} to pulmonary capillary blood volume (Vc) ratio (DM_{CO}/Vc), as they relate to the percentage of maximal alveolar volume (V_A) (*x*-axis) compared to their percentage value at maximal V_A (*y*-axis) and (**b**) rates of alveolar uptake for NO and CO per unit time and pressure, K_{NO} and K_{CO} (mathematically equivalent to DL_{NO}/V_A and DL_{CO}/V_A,respectively) and DM and Vc, both per unit V_A (DM/V_A and Vc/V_A), as the expansion of the lung is changed voluntarily in normal subjects (100% of maximal V_A, which is approximately TLC, and 50% of maximal V_A, which is approximately FRC). Note in (**a**) that with diminishing lung expansion (ΔV_A), ΔDL_{NO} is better related to DM_{CO} (ΔDM_{CO}) and ΔDM_{CO}/Vc change than the DL_{CO} change. In (**b**), ΔK_{CO} is a better reflection of changes in the pulmonary microcirculation (Vc/V_A) than the K_{NO}; decrease of DM/V_A with ΔV_A suggests *isotropic change* as alveolar dimensions reduce with concomitant thickening of the alveolar-capillary membrane. Reproduced with permission of the © ERS 2018: *European Respiratory Journal Feb 2017, 49 (2) 1600962; DOI:* https://doi.org/10.1183/13993003.00962-2016

General recommendations. For each parameter, the 90% confidence interval represents the normal range. Fixed ratios, such as FEV_1/VC of 0.70 and 80% of predicted, should not be assumed as lower limits of normal, because they are age- and sex-biased, thus causing underdiagnosis in young and female subjects but overdiagnosis in old and male subjects.

Key Points

- Static lung volumes are regulated by the elastic properties of the lung and chest wall.
- Static lung hyperinflation is defined as an increase in relaxation volume of the respiratory system, generally due to loss of lung elastic recoil (emphysema).
- Dynamic lung hyperinflation is defined as an end-tidal expiratory volume above the relaxation volume and is generally due to expiratory flow limitation.
- Expiratory flow limitation occurs during forced expiration in both healthy and diseased subjects, though at different levels.
- Spirometry generally allows detecting obstructive disorders, whereas measurements of lung volumes are necessary to confirm lung restriction.
- Forced inspiratory flow can help identify extrathoracic or fixed airflow obstruction.
- No parameter from forced expiratory manoeuvre is specific for small airway function.
- Ventilation inhomogeneity is present in normal lungs but is much greater in lung disease.
- The uptake of carbon monoxide is largely dependent on pulmonary microcirculation.
- Lung diffusing capacity is the result of various combinations of carbon monoxide uptake, and alveolar volume; thus correction for lung volume is not reliable.

References

1. Agostoni E, Hyatt RE. Static behaviour of the respiratory system. In: Macklem PT, Mead J, editors. Handbook of physiology, The respiratory system. Mechanics of breathing. Section 3, part 1, vol. III. Bethesda: American Physiological Society; 1986. p. 113–30.
2. Loring SH, O'Donnell CR, Butler JP, et al. Transpulmonary pressures and lung mechanics with glossopharyngeal insufflation and exsufflation beyond normal lung volumes in competitive breath-hold divers. J Appl Physiol. 2007;102:841–6.
3. Seccombe LM, Rogers PG, Mai N, et al. Features of glossopharyngeal breathing in breath-hold divers. J Appl Physiol. 2006;101:799–801.
4. Gibson GJ, Pride NB, O'cain C, et al. Sex and age differences in pulmonary mechanics in normal nonsmoking subjects. J Appl Physiol. 1976;41:20–5.

5. Younes M, Kivinen G. Respiratory mechanics and breathing pattern during and following maximal exercise. J Appl Physiol. 1984;57:1773–82.
6. Woolcock AJ, Read J. Lung volumes in exacerbations of asthma. Am J Med. 1966;41:259–73.
7. Pride NB, Macklem PT. Lung mechanics in disease. In: Macklem PT, Mead J, editors. Handbook of physiology, The respiratory system. Mechanics of breathing. Section 3, part 2, vol. III. Bethesda: American Physiological Society; 1986. p. 659–92.
8. Leith DE, Mead J. Mechanisms determining residual volume of the lungs in normal subjects. J Appl Physiol. 1967;23:221–7.
9. Hutchinson J. On the capacity of the lungs, and on the respiratory movements, with the view of establishing a precise and easy method of detecting disease by the spirometer. Lancet. 1846;1:630–2.
10. Aaron SD, Dales RE, Cardinal P. How accurate is spirometry at predicting restrictive pulmonary impairment? Chest. 1999;115:869–73.
11. Vinegar A, Sinnett EE, Leith DE. Dynamic mechanisms determine functional residual capacity in mice, Mus musculus. J Appl Physiol. 1979;46:867–71.
12. Pellegrino R, Brusasco V, Rodarte JR, et al. Expiratory flow limitation and regulation of end-expiratory lung volume during exercise. J Appl Physiol. 1993;74:2552–8.
13. Pellegrino R, Violante B, Nava S, et al. Relationship between expiratory airflow limitation and hyperinflation during methacholine-induced bronchoconstriction. J Appl Physiol. 1993;75:1720–7.
14. Duranti R, Filippelli M, Bianchi R, et al. Inspiratory capacity and decrease in lung hyperinflation with albuterol in COPD. Chest. 2002;122:2009–14.
15. Pepe PE, Marini JJ. Occult positive end-expiratory pressure in mechanically ventilated patients with airflow obstruction: the auto-PEEP effect. Am Rev Respir Dis. 1982;126:166–70.
16. Anthonisen NR. Tests of mechanical function. In: Macklem PT, Mead J, editors. Handbook of physiology, The respiratory system. Mechanics of breathing. Section 3, part 2, vol. III. Bethesda: American Physiological Society; 1986. p. 753–84.
17. DuBois AB, Botelho SY, Comroe JH Jr. A new method for measuring airway resistance in man using a body plethysmograph: values in normal subjects and in patients with respiratory disease. J Clin Invest. 1956;35:327–35.
18. Miller RD, Offord KP. Roentgenologic determination of total lung capacity. Mayo Clin Proc. 1980;55:694–9.
19. Krayer S, Rehder K, Beck KC, et al. Quantification of thoracic volumes by three-dimensional imaging. J Appl Physiol. 1987;62:591–8.
20. Rodenstein DO, Stănescu DC. Reassessment of lung volume measurement by helium dilution and by body plethysmography in chronic air-flow obstruction. Am Rev Respir Dis. 1982;126:1040–4.
21. O'Donnell CR, Bankier AA, Stiebellehner L, et al. Comparison of plethysmographic and helium dilution lung volumes: which is best for COPD? Chest. 2010;137:1108–15.
22. Shore SA, Huk O, Mannix S, et al. Effect of panting frequency on the plethysmographic determination of thoracic gas volume in chronic obstructive pulmonary disease. Am Rev Respir Dis. 1983;128:54–9.
23. Rodarte JR, Rehder K. Dynamics of respiration. In: Macklem PT, Mead J, editors. Handbook of physiology, The respiratory system. Mechanics of breathing. Section 3, part 1, vol. III. Bethesda: American Physiological Society; 1986. p. 131–44.
24. Peslin R, Fredberg JJ. Oscillation mechanics of the respiratory system. In: Macklem PT, Mead J, editors. Handbook of physiology, The respiratory system. Mechanics of breathing. Section 3, part 1, vol. III. Bethesda: American Physiological Society; 1986. p. 145–77.
25. Peslin R, Felicio da Silva J, et al. Respiratory mechanics studied by forced oscillations during artificial ventilation. Eur Respir J. 1993;6:772–84.
26. Dellacà RL, Santus P, Aliverti A, et al. Detection of expiratory flow limitation in COPD using the forced oscillation technique. Eur Respir J. 2004;23:232–40.
27. Hyatt RE, Wilcox RE. Extrathoracic airway resistance in man. J Appl Physiol. 1961;16:326–30.

28. Mead J, Whittenberger JL. Evaluation of airway interruption technique as a method for measuring pulmonary air-flow resistance. J Appl Physiol. 1954;6:408–16.
29. Yanai M, Sekizawa K, Ohrui T, et al. Site of airway obstruction in pulmonary disease: direct measurement of intrabronchial pressure. J Appl Physiol. 1992;72:1016–23.
30. Brusasco V, Warner DO, Beck KC, et al. Partitioning of pulmonary resistance in dogs: effect of tidal volume and frequency. J Appl Physiol. 1989;66:1190–6.
31. Tiffeneau R, Pinelli A. Régulation bronchique de la ventilation pulmonaire. J Fr Med Chir Thorac. 1948;2:221–44.
32. Hyatt RE, Schilder DP, Fry DL. Relationship between maximum expiratory flow and degree of lung inflation. J Appl Physiol. 1958;13:331–6.
33. Hyatt RE. Forced expiration. In: Macklem PT, Mead J, editors. Handbook of physiology, The respiratory system. Mechanics of breathing, Section 3, part 1, vol. III. Bethesda: American Physiological Society; 1986. p. 295–314.
34. Permutt S, Riley RL. Hemodynamics of collapsible vessels with tone: the vascular waterfall. J Appl Physiol. 1963;18:924–32.
35. Mead J, Turner JM, Macklem PT, et al. Significance of the relationship between lung recoil and maximum expiratory flow. J Appl Physiol. 1967;22:95–108.
36. Dawson SV, Elliott EA. Wave-speed limitation on expiratory flow – a unifying concept. J Appl Physiol. 1977;43:498–515.
37. Quanjer PH, Weiner DJ, Pretto JJ, et al. Measurement of FEF 25–75% and FEF 75% does not contribute to clinical decision making. Eur Respir J. 2014;43:1051–8.
38. Schilder DP, Roberts A, Fry DL. Effects of gas density and viscosity on the maximal expiratory flow-volume relationships. J Clin Invest. 1963;42:1705–13.
39. Meadows JA 3rd, Rodarte JR, Hyatt RE. Density dependence of maximal expiratory flow in chronic obstructive pulmonary disease. Am Rev Respir Dis. 1980;121:47–53.
40. Davis J, Schroter RC, Sudlow MF. The investigation of the effects of gas density and viscosity on the physics of flow in small airways during forced expiration. J Physiol. 1972;223:9P–10P.
41. Nadel JA, Thierney DF. Effect of a previous deep inspiration on airway resistance in man. J Appl Physiol. 1961;16:717–9.
42. Lim TK, Ang SM, Rossing TH, et al. The effects of deep inhalation on maximal expiratory flow during intensive treatment of spontaneous asthmatic episodes. Am Rev Respir Dis. 1989;140:340–3.
43. Fairshter RD. Effect of a deep inspiration on expiratory flow in normals and patients with chronic obstructive pulmonary disease. Bull Eur Physiopathol Respir. 1986;22:119–25.
44. Ingram RH Jr, Schilder DP. Effect of gas compression on pulmonary pressure, flow, and volume relationship. J Appl Physiol. 1966;21:1821–6.
45. Pellegrino R, Antonelli A, Crimi E, et al. Dependence of bronchoconstrictor and bronchodilator responses on thoracic gas compression volume. Respirology. 2014;19:1040–5.
46. Pellegrino R, Brusasco V. Lung hyperinflation and flow limitation in chronic airway obstruction. Eur Respir J. 1997;10:543–9.
47. Ninane V, Leduc D, Kafi SA, et al. Detection of expiratory flow limitation by manual compression of the abdominal wall. Am J Respir Crit Care Med. 2001;163:1326–30.
48. Koulouris NG, Valta P, Lavoie A, et al. A simple method to detect expiratory flow limitation during spontaneous breathing. Eur Respir J. 1995;8:306–13.
49. Hage R, Aerts JGJV, Verbraak AFM, et al. Detection of flow limitation during tidal breathing by the interruptor technique. Eur Resp J. 1995;8(11):1910–4.
50. Miller RD, Hyatt RE. Obstructing lesions of the larynx and trachea: clinical and physiologic characteristics. Mayo Clin Proc. 1969;44:145–61.
51. Storey WF, Staub NC. Ventilation of terminal air units. J Appl Physiol. 1962;17:391–7.
52. Engel LA, Anthonisen NR. Demonstration of airway closure in man. J Appl Physiol. 1975;38:1117–25.
53. Fowler WS. Lung function studies. III. Uneven pulmonary ventilation on normal subjects and in patients with pulmonary disease. J Appl Physiol. 1949;2:283–99.

54. Comroe JH Jr, Fowler WS. Lung function studies. VI Detection of uneven alveolar ventilation during a single breath of oxygen. Am J Med. 1951;10:408–13.
55. Prisk GK. Microgravity and the lung. J Appl Physiol. 2000;89:385–96.
56. Cosio M, Ghezzo H, Hogg JC, et al. The relations between structural changes in small airways and pulmonary function tests. N Engl J Med. 1978;298:1277–81.
57. Milic-Emili J, Torchio R, D'Angelo E. Closing volume: a reappraisal (1967–2007). Eur J Appl Physiol. 2007;99:567–83.
58. Darling RC, Cournand A, Mansfield JS, et al. Studies on the intrapulmonary mixture of gases. I. Nitrogen elimination from blood and body tissues during high oxygen breathing. J Clin Invest. 1940;19:591–7.
59. Becklake MR. A new index of the intrapulmonary mixture of inspired air. Thorax. 1952;7:111–6.
60. Bouhuys A. Pulmonary nitrogen clearance in relation to age in healthy males. J Appl Physiol. 1963;18:297–300.
61. Oude Engberink E, Ratjen F, Davis SD, et al. Inter-test reproducibility of the lung clearance index measured by multiple breath washout. Eur Respir J. 2017;5(50):1700433.
62. Gustafsson PM, Aurora P, Lindblad A. Evaluation of ventilation maldistribution as an early indicator of lung disease in children with cystic fibrosis. Eur Respir J. 2003;22:972–9.
63. Crawford AB, Makowska M, Paiva M, et al. Convection- and diffusion-dependent ventilation maldistribution in normal subjects. J Appl Physiol. 1985;59:838–46.
64. Verbanck S, Paiva M. Gas mixing in the airways and airspaces. Compr Physiol. 2011;1:809–34.
65. Weibel ER, Federspiel WJ, Fryder-Doffey F, et al. Morphometric model for pulmonary diffusing capacity. I. Membrane diffusing capacity. Respir Physiol. 1993;93:125–49.
66. Scheid P, Piiper J. Diffusion. In: Crystal RG, West JB, Weibel ER, Barnes PJ, editors. The lung: scientific foundations. Philadelphia, PA: Lippincott-Raven; 1997. p. 1681–071.
67. Hsia CC, Wagner PD, Dane DM, et al. Predicting diffusive alveolar oxygen transfer from carbon monoxide-diffusing capacity in exercising foxhounds. J Appl Physiol. 2008;105:1441–7.
68. Cotes JE, Dabbs JM, Elwood PC, et al. Iron-deficiency anemia: its effect on transfer factor for the lung (diffusing capacity) and ventilation and cardiac frequency during submaximal exercise. Clin Sci. 1972;42:325–35.
69. Krogh M. The diffusion of gases through the lungs of man. J Physiol. 1915;49:271–300.
70. Blakemore WS, Forster RE, Morton JW, et al. A standardized breath holding technique for the clinical measurement of the diffusing capacity of the lung for carbon monoxide. J Clin Invest. 1957;36:1–17.
71. Graham BL, Brusasco V, Burgos F, et al. 2017 ERS/ATS standards for single-breath carbon monoxide uptake in the lung. Eur Respir J. 2017;1600016:49.
72. Stanojevic S, Graham BL, Cooper BG, et al. Official ERS technical standards: global lung function initiative reference values for the carbon monoxide transfer factor for Caucasians. Eur Respir J. 2017;50:1700010.
73. Hughes JM, Pride NB. In defence of the carbon monoxide transfer coefficient Kco (TL/VA). Eur Respir J. 2001;17:168–74.
74. Roughton FJ, Forster RE. Relative importance of diffusion and chemical reaction rates in determining the rate of exchange of gases in the human lung, with special reference to true diffusing capacity of pulmonary membrane and volume of blood in the lung capillaries. J Appl Physiol. 1957;11:290–302.
75. Hsia CC, McBrayer DG, Ramanathan M. Reference values of pulmonary diffusing capacity during exercise by a rebreathing technique. Am J Respir Crit Care Med. 1995;152:658–65.
76. Carlsen E, Comroe JH Jr. The rate of uptake of carbon monoxide and of nitric oxide by normal human erythrocytes and experimentally produced spherocytes. J Gen Physiol. 1958;42:83–107.
77. Guénard H, Varène N, Vaida P. Determination of lung capillary blood volume and membrane diffusing capacity by measurement of NO and CO transfer. Respir Physiol. 1987;70:113–20.
78. Borland CD, Higenbottam TW. A simultaneous single breath measurement of pulmonary diffusing capacity with nitric oxide and carbon monoxide. Eur Respir J. 1989;2:56–63.

79. Borland C, Bottrill F, Jones A, et al. The significant blood resistance to lung nitric oxide transfer lies within the red cell. J Appl Physiol. 2014;116:32–41.
80. Barisione G, Brusasco C, Garlaschi A, et al. Lung diffusing capacity for nitric oxide as a marker of fibrotic changes in idiopathic interstitial pneumonias. J Appl Physiol. 2016;120:1029–38.
81. Pellegrino R, Viegi G, Brusasco V, et al. Interpretative strategies for lung function tests. Eur Respir J. 2005;26:948–68.
82. Stam H, Hrachovina V, Stijnen T, et al. Diffusing capacity dependent on lung volume and age in normal subjects. J Appl Physiol. 1994;76:2356–63.

Part III

Respiratory Diseases and Exercise

Asthma

4

Matteo Bonini

Abstract

Extensive evidence exists on the beneficial effect of training and rehabilitation programs in asthma. On the other hand, intense and repeated physical exercise may trigger transient airway narrowing, defined exercise-induced bronchoconstriction (EIB). The prevalence of EIB has been reported to be up to 90% in asthmatic patients, reflecting the level of disease control. However, EIB may develop even in subjects without clinical asthma, particularly in athletes, children and subjects with atopy or rhinitis and following respiratory infections. The occurrence of EIB, however, can be optimally managed and should not prevent from an adequate practice of physical activity.

4.1 Introduction

Regular physical activity is strongly recommended by worldwide healthcare systems and evidence-based guidelines as one of the most effective tools to prevent chronic diseases and maintain good health [1]. Indeed, extensive evidence exists on the beneficial effect of training and rehabilitation programs in respiratory diseases, particularly in asthma [2]. It has been in fact shown that physical activity improves symptoms, quality of life, exercise capacity and pulmonary function, as well as reduces airway inflammation and responsiveness in asthmatic subjects [3–5].

M. Bonini (✉)
National Heart and Lung Institute (NHLI), Royal Brompton Hospital and Imperial College London, London, UK

Department of Cardiovascular and Thoracic Sciences, Fondazione Policlinico Universitario A. Gemelli–IRCCS, Rome, Italy
e-mail: m.bonini@imperial.ac.uk

A. Cogo et al. (eds.), *Exercise and Sports Pulmonology*,
https://doi.org/10.1007/978-3-030-05258-4_4

On the other hand, intense physical exercise may trigger airway narrowing by imposing high demands on the respiratory system, requiring subjects to ventilate primarily through the mouth and bypass the nasal filter, with a subsequent increased pulmonary exposure to inhalant allergens, pollutants, irritants and adverse (i.e. cold, dry) environmental conditions [6]. Furthermore, intense physical training may induce a transient status of immune downregulation, clinically associated with an increased prevalence of atopy and viral upper respiratory tract infections (URTI)—see Chap. 9—both representing relevant risk factors for the onset and worsening of asthma [7, 8].

The transient airway narrowing that occurs as a result of exercise is defined exercise-induced bronchoconstriction (EIB) [6]. Already in the first century A.D., Aretaeus the Cappadocian described respiratory symptoms induced by physical exercise: "if from running, gymnastics, or any other work, breathing becomes difficult, it is called asthma" [9]. However, a scientific objective interest for this phenomenon can be dated back to 1960, when Jones and co-workers focused on the physiologic response to exercise in asthmatic children and named the airway obstruction after an exercise challenge "exercise-induced asthma" (EIA) [10]. Subsequent studies defined the different patterns of response to exercise in asthmatic patients, as well as the influence of antiasthmatic drugs on EIA [11, 12]. Although exercise may trigger bronchial obstruction and respiratory symptoms in almost all asthmatic patients, independently from the underlying causes and mechanisms of asthma [13], some authors consider EIA a distinct phenotype of asthma [14]. However, the concept that exercise may induce bronchial obstruction only in asthmatic patients is at present under debate [15]. In fact, despite the physiologic response to exercise usually results in slight bronchodilation, EIB may develop even in subjects without clinical asthma [9]. To bring some clarity to this controversial issue, a practice parameter, jointly developed by the American Academy/College of Allergy Asthma and Immunology (AAAAI/ACAAI) [16], recommended to abandon the term EIA, and more recently an American Thoracic Society Clinical Practice Guideline [6] suggested to name EIB with asthma (EIBa), the occurrence of bronchial obstruction after exercise in asthmatic patients, and EIB without asthma (EIBwa), the occurrence of exercise-induced bronchoconstriction in subjects without other symptoms and signs of clinical asthma.

4.2 Clinical Features

EIB typically develops within 15 min following at least 5–8 min of high-intensity aerobic training (>85% of maximal voluntary ventilation), although it can also occur during exercise, and spontaneously resolves within 60 min [17]. After an episode of EIB, there is often a refractory period of about 1–3 h during which, if exercise is repeated, the bronchoconstriction is less accentuated [12]. Most common symptoms include cough, dyspnoea, breathlessness, wheezing and chest tightness [6].

4.3 Prevalence

The prevalence of EIB varies from 5 to 20% in the general population and has been reported to be up to 90% in asthmatic subjects, reflecting the level of disease control, with EIBa occurring more frequently in more severe and uncontrolled asthmatic patients [13]. EIBwa is also particularly frequent in athletes [18], children [19, 20] and subjects with rhinitis [21] and following respiratory infections [22].

In particular, several studies called attention to an increased occurrence of asthma and EIB in athletes, with prevalence rates widely ranging from 3.7 to 54.8% (Table 4.1) depending on the study population and the criteria used for diagnosis (i.e. questionnaires, anti-doping records, baseline spirometry, bronchial provocation challenges). Independently from these potential confounders, studies performed in comparable samples and with similar diagnostic methodologies seem to indicate that the asthma incidence is on the increase: from 9.7% in 1976 to 11.2% in 1984, 16.7% in 1996 and 21.0% in 2000 in the US Olympic delegation [23–25]. More recently, a 12-year study including four cross-sectional surveys performed between 2000 and 2012, before Summer and Winter Olympics, showed that the prevalence of asthma in 659 Italian Olympic athletes was 14.7%, with a significant increase from 2000 (11.3%) to 2008 (17.2%) [26].

With regard to a gender effect [27], a study recently performed in 187 elite athletes (101 swimmers and 86 tennis players) showed a higher prevalence of asthma symptoms in females, although there was no significant difference in the prevalence of EIB when measured through a mannitol and a sport-specific challenge [28]. Norqvist et al. also reported that, compared to males, elite female athletes had a higher prevalence of asthma, respiratory symptoms, use of medications and healthcare services [29].

It has been also extensively reported that asthma and allergic rhinitis frequently coexist, with symptoms of rhinitis being reported in 80–90% of asthma patients and asthma symptoms reported in 20–40% of patients with allergic rhinitis [21]. Prospective studies also suggest that rhinitis frequently precedes the development of asthma [30] and that many patients with rhinitis alone show non-specific bronchial hyperresponsiveness after exercise or methacholine, this being a risk factor for developing asthma [31]. Furthermore, it has been proven that the severity of allergic rhinitis and asthma are related and that proper management of allergic rhinitis improves asthma control [21]. Additionally, exercise can be a trigger for rhinitis, especially in outdoor sports and even greater with cold dry air exposure in winter sports, e.g. the "skier's nose" [32]. On the basis of all the above, the ARIA recommendation [21] to screen every subject with rhinitis for asthma should be also extended to athletes [33].

Despite extensive epidemiological data, EIB progression in athletes has not been yet fully studied. However, in a 5-year prospective study, subjects who stopped training experienced an attenuation or in some circumstances disappearance of

Table 4.1 Prevalence of asthma and EIB among athletes

Study population (n)	Prevalence	Methodology for diagnosis	Reference
US College athletes (80)	42.5%	Questionnaire, exercise challenge	Burnett DM, 2016
US 1998 Olympic team (170)	23.0%	Spirometry, exercise challenge	Wilber RL, 2000
US 1998 Olympic team (196)	21.9%	Questionnaire	Weiler JM, 2000
Summer athletes (162)	22.8%	Questionnaire, spirometry, histamine challenge	Helenius IJ, 1998
Australian 2000 Olympic team (214)	21.0%	Questionnaire	Katelaris CH, 2000
US 1996 Olympic team (699)	16.7%	Questionnaire	Weiler JM, 1998
Italian Olympicathletes (659)	14.7%	Questionnaires, lung function tests	Bonini M, 2015
Polish 2008 Olympic team (222)	11.3%	Questionnaire, spirometry, methacholine challenge	Kurowski M, 2016
Italian 2000 pre-Olympic team (265)	10.9%	Questionnaire, spirometry	Lapucci G, 2003
Australian 1976 Olympic team (185)	9.7%	Physical examination	Fitch KD, 1984
Australian 1980 Olympic team (106)	8.5%	Physical examination	Fitch KD, 1984
Spanish 1982 Olympic team (495)	4.4%	Questionnaire	Drobnic F, 1994
US 1984 Olympic team (597)	4.3%	Questionnaire, exercise challenge	Voy RO, 1984
Swiss athletes (2060)	3.7%	Questionnaire	Helbling A, 1990
Cross-country skiers (42)	54.8%	Questionnaire, spirometry, methacholine challenge	Larsson K, 1993
Swedish and Norwegian cross-country skiers (171)	42.0%/12.0%	Questionnaire, spirometry, methacholine challenge	Sue-Chu M, 1996
Ice hockey players (88)	21.5%	Questionnaire, spirometry, histamine challenge	Lumme A, 2003
Ice hockey players (50)	11.5%	Questionnaire, spirometry, methacholine and exercise challenge	Leuppi JD, 1998
Crosscountry skiers (20)	10.0%	Exercise challenge	Pohjantähti H, 2005
Swimmers (90)	39.0%	EVH challenge	Bougault V, 2010
US swimmers (738)	13.4%	Questionnaire	Potts J, 1996
Marathon runners (208)	32.0%	Questionnaire	Robson-Ansley P, 2012
Finnish runners (103)	15.5%	Questionnaire	Tikkanen H, 1994
US track and field (73)	15.1%	Exercise challenge	Schoene RB, 1997
Figure skaters (124)	35.0%	Exercise challenge	Mannix ET, 1996
US football players (156)	11.5%	Questionnaire, methacholine challenge	Weiler JM, 1986

Blue = various sports; green = winter sports; purple = swimming; orange = track and field; yellow = others

From Bonini M, Silvers W. Immunol Allergy Clin North Am. 2018;38(2):205–214

EIB, whereas bronchial responsiveness, exercise-induced respiratory symptoms and eosinophilic airway inflammation increased among those who continued strenuous physical exercise, regardless of the pharmacological treatment strategies [34]. Put into context, ongoing intense training therefore appears to be a causative, and not just a concomitant, factor of airway inflammation and narrowing.

4.4 Pathogenesis

The increase in airways osmolarity due to respiratory water loss and the vasodilation associated with airways rewarming have been commonly reported to be the major determinants of EIB (osmotic and thermal theories) [35, 36]. However, as with the current approach in asthma phenotypes, different endotypes (i.e. disease subtypes specifically defined by functional or pathological molecular mechanisms and/or treatment response) of EIB have been recently reported (Fig. 4.1) [37]. Intense physical training may induce a transient status of immune downregulation with a shift towards a relatively prevalent T2-high response, clinically associated with an increased prevalence of atopy [8, 26]. However, despite in a large proportion of EIB subjects the occurrence of bronchial obstruction following exercise is associated with allergic diseases [26] and markers of a T2-high response [38], EIB is also present in subjects with no evidence of atopic sensitisation. In such non-T2-high variants of EIB, the bronchial epithelial damage directly caused by physical activity has been suggested as a relevant pathogenic mechanism [37]. A direct injury of the bronchial epithelium might also be caused by viral upper respiratory tract infections, which are reported to occur more frequently in athletes [39]. Inflammatory mediators released by the damaged epithelium have been found both in sputum (i.e. interleukin-8) [40] and in serum (i.e. nerve growth factor) [41] and seem to be associated with a neutrophilic or mixed neutrophilic/eosinophilic inflammatory response. Furthermore, the evidence that CC16 proteins, secreted by club cells in the distal bronchioles to protect the respiratory tract against oxidative stress and inflammation, are increased in urine and serum following an exercise challenge

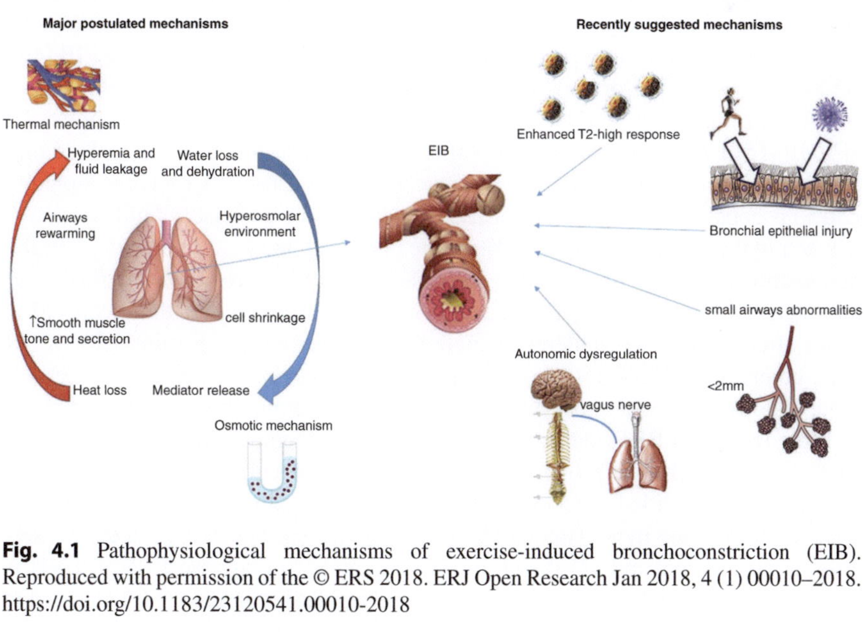

Fig. 4.1 Pathophysiological mechanisms of exercise-induced bronchoconstriction (EIB). Reproduced with permission of the © ERS 2018. ERJ Open Research Jan 2018, 4 (1) 00010–2018. https://doi.org/10.1183/23120541.00010-2018

[42] has prompted renewed and increasing interest in the role of small airways in EIB. Notably, small-airway abnormalities have been related to the onset and severity of asthma and may occur even in the absence of a response in the large airways [43]. In addition, an alternative noninflammatory mechanism, autonomic dysregulation with an enhanced parasympathetic response, measured by both pupillometry and heart rate variability, has been shown to significantly relate to EIB [44].

4.5 Diagnosis

A careful history taking and physical examination is always recommended. The use of questionnaires specifically developed and validated for screening atopy and allergic diseases in athletes may represent an additional useful and easy-to-use diagnostic tool [45]. However, research performed over the past years has consistently revealed a poor relationship between the presence of "asthma-like" symptoms and objective evidence of EIB [46].

Furthermore, baseline spirometry appears to be poorly predictive of EIB in athletes, often being within the normal ranges even in the presence of disease [47]. The assessment of small airways, whose involvement has been reported in EIB [48], through ad hoc diagnostic investigations such as impulse oscillometry, multibreath nitrogen washout and fraction exhaled nitric oxide (FeNO) at multi-flows, might provide further useful information, particularly in the early and mild forms [43]. However, in order to establish a secure diagnosis of EIB, it is strongly recommended to perform objective bronchoprovocative tests (BPTs) to document dynamic changes in airway function.

Measuring the change in the forced expiratory volume in 1 s (FEV1) before and after a standardized exercise challenge test (ECT), in the laboratory or in the field, represents the most intuitive and commonly adopted approach to diagnose EIB [49]. An ECT should be performed in subjects with EIBa only when their baseline FEV1 is $\geq$70% of normal. A $\geq$ 10% fall in FEV1 at any two consecutive time points (1, 3, 5, 10, 15, 20, 25, 30, 60, 90 min.) after 6 to 8 min of treadmill or cycloergometer exercise in laboratory-specific ambient conditions (20–25 °C; relative humidity <50%) is considered diagnostic of EIB. The intensity of exercise should be enough to reach in the first 2–3 min. 40–60% of the predicted maximum voluntary ventilation (estimated as baseline FEV1 $\times$ 35) or 80–90% of the predicted maximal heart rate (calculated by 220—age). Indeed, it has been reported that the mean fall in FEV1 after an exercise challenge is more than doubled after achieving 95% of HR max, compared to 85% [50]. Sports-specific challenges in the field may be also used, although these are more difficult to standardize, limiting their application.

Moreover, other BPTs can be adopted as surrogate diagnostic tools for EIB. Direct BPTs (i.e. methacholine provocation) are accurate to document bronchial hyperreactivity in EIBa, while indirect tests, such as eucapnic voluntary hyperpnea (EVH), mannitol and saline hyperosmolar challenges, better reproduce the effects of exercise on the airways and are therefore more accurate to diagnose EIBwa. EVH is often preferred in highly trained athletes, where standard criteria for ECT may fail to

Table 4.2 Thresholds set by the International Olympic Committee (IOC) to document EIB through the different bronchial provocation tests (BPTs)

Diagnostic procedure	Diagnostic criteria
Exercise challenge	↓ FEV1 ≥ 10%
Methacholine challenge	↓ FEV1 ≥ 20% with a: PC20 ≤ 4 mg/mL (for subjects not taking ICS) or PC20 ≤ 16 mg/mL (for subjects taking ICS for at least 1 month)
Eucapnic voluntary hyperpnea (EVH)	↓ FEV1 ≥ 10%
Hyperosmolar tests (mannitol, saline)	↓FEV1 ≥ 15%

reproduce the bronchoprovocative stimulus experienced while practising their own sport disciplines [51]. However, correlations between ECT and other indirect BPTs are at present arguable. Thresholds set by the International Olympic Committee (IOC) for a positive response to the different BPTs are reported in Table 4.2.

The differential diagnosis of EIB should take into account physiologic limitations; anxiety; exercise-induced laryngeal dysfunctions, hyperventilation and hypoxemia; dyspnoea on exertion in obese or poorly fit individuals; shortness of breath with exercise due to lung diseases other than asthma and cardiac diseases; anaemia; and myopathies [6]. In particular, vocal cord dysfunction (VCD) is increasingly recognized as a condition that may mimic EIB. However, in VCD, the inspiratory stridor during exercise usually resolves within 5 min representing the major differential sign, associated with negative BPT result and poor response to antiasthmatic drugs. Vocal cord dysfunction may also coexist with EIB. If pruritus, urticaria or systemic reactions are associated with symptoms of EIB, the diagnosis of exercise-induced urticaria or anaphylaxis should be at last considered [52].

4.6 Management

Treatment of both EIBa and EIBwa is essentially based on reversing bronchial obstruction by using short-acting beta-2 agonists [13].

4.6.1 Non-pharmacological Prevention

Similar measures can be adopted in both EIBa and EIBwa. These include, whenever possible, avoiding exercise in an at-risk air environment because of temperature, humidity, pollutants and specific allergens in sensitized subjects [16]. Education about self-management is essential and should include advice about environmental measures, inhaler technique and the use of an action plans, in addition to regular follow-up [13]. Progressive warming-up and cooling-down periods are constantly suggested [53]. Some athletes may also take advantage of the refractory period following bronchial obstruction deliberately induced by hyperventilation or by an intense exercise challenge. The use of face masks to warm and humidify the air has been reported to provide benefits, especially in winter athletes [54]. There is at last some evidence that weight loss and dietary factors, such as vitamin D supplementation, may be helpful in reducing the risk and severity of EIB [55, 56]. The potential occurrence of EIB should not prevent subjects from an adequate practice of physical exercise, which has been proven not to be associated with an increased risk of asthma developing or worsening [4], and should instead represent part of their treatment.

4.6.2 Pharmacological Prevention

Because EIBa is a sign of poor asthma control, prevention essentially consists of following international guidelines to avoid symptoms and reduce the risk of exacerbations [13]. Multiple therapeutic options seem also appropriate to prevent EIBwa, although usually they do not completely avoid the occurrence of bronchoconstriction, but rather attenuate it or shift the dose-response relationship, so that some submaximal efforts become tolerated.

Regular use of inhaled corticosteroid (ICS) represents a key strategy for controlling asthma and therefore is a recommended treatment to prevent EIBa [57]. The prophylactic administration of ICS has been also suggested in EIBwa, particularly if physical activity is performed regularly (>3 times per week), representing a repetitive stimulus for the onset of bronchoconstriction [16]. However, the use of ICS in the prevention of EIBwa may be controversial. In fact this pharmacological strategy is at present not supported by ad hoc designed clinical trials, and response to treatment may be impaired in subjects with underlying non-eosinophilic inflammatory pattern.

Beta-2 adrenergic drugs, both short- and long-acting (SABA and LABA), when given in a single inhaled dose or with intermittent administration before exercise, are the most effective drugs to prevent both EIBa and EIBwa [58], providing complete protection against exercise (FEV1 fall <10%) in approx. 70% of subjects [59]. The effect usually lasts 2 to 4 h for SABA and up to 12 h for LABA. Heterogeneity observed in the efficacy of beta-2 adrenergic agents to prevent EIB is not dependent on the type of molecule used, but rather on the population sample studied, with more variable effects reported in children [59]. However, the chronic use of SABA and LABA often results in a reduction of the duration and/or magnitude of protection against EIB with cross-reacting tolerance to other beta-2 agonists [59, 60]. This impaired efficacy has been shown to be predicted by baseline levels of FeNO [61]. Salpeter and co-authors reported that tolerance to beta-2 agonists is only partially prevented by concomitant use of ICS [61]. Furthermore, daily use of SABA and LABA may result even in a worsening of EIB [62] and expose subjects to an increased risk of cardiovascular side effects and death [63, 64]. Therefore, SABA and LABA should be used with caution on a regular basis to prevent EIB. LABA administration should be also always avoided without concomitant use of ICS according to the US Food and Drug Administration (FDA) warning.

Mast cell stabilizers, disodium cromoglycate and nedocromil sodium, attenuate both EIBa and EIBwa when inhaled shortly before exercise but have a short duration of action [65].

Leukotriene antagonists (i.e. montelukast) have been reported to be effective in preventing EIBa. However, protection occurs in approximately 50% of subjects and may not be complete [66].

Ipratropium bromide prevents EIBa, although this effect is not consistent among patients and may be variable in the same patient [67]. Whether subjects with EIBwa or with a prevalent autonomic imbalance represent an EIB phenotype more responsive to anticholinergic agents represents an interesting hypothesis, still waiting for further experimental testing [68].

Calcium channel blockers, beta-adrenergic receptor antagonists, inhaled furosemide, heparin and hyaluronic acid have been studied to prevent EIB with inconsistent results.

Special precautions must be taken, at last, with respect to the World Anti-Doping Agency (WADA) rules on the use of EIB medications in competitive athletes—see Chap. 13.

4.7 Tips and Pitfalls

The intensity, duration and type of training have been associated with the occurrence of bronchial symptoms, airway hyperresponsiveness and asthma in elite athletes (Table 4.3).

Asthma is most commonly found in athletes performing endurance activities, such as long-distance running, cycling, triathlon and pentathlon. The high prevalence of EIB among endurance athletes has been mainly attributed to an increased minute ventilation through the mouth (bypassing the nasal filter) and exposure to allergens and pollutants. In major national and international competitions, local pollen counts (i.e. www.polleninfo.org) and air quality forecasts should be therefore always made available in advance to athletes, their coaches and medical teams.

Environmental factors (i.e. cold and dry air) also play a relevant role for athletes practising winter sports.

Swimming has been long considered a safe and recommended sport activity for subjects with asthma due to the inhalation of humid air; however, despite conflicting data, an increased risk of EIB with swimming and pool attendance has been reported [34, 69]. These findings are thought to be the result of repeated hyperventilation challenges together with the exposure to chlorine-based derivatives, commonly used to disinfect swimming pools, such as trichloramine. This hypothesis is further supported by studies on occupational asthma in swimming pool workers and lifeguards and by studies comparing exposures to non-chlorinated pools (copper-silver pools) vs. chlorinated pools.

Table 4.3 Sport disciplines and risk of exercise-induced bronchoconstriction (EIB)

Low-risk sports	Medium-risk sports	High-risk sports
All sports where the exercise lasts <5–8 min	Team sports where continuous exercise rarely lasts more than 5–8 min	All sports where the exercise lasts >5–8 min and/or is performed in special environments (i.e. dry/cold air, chlorinated pools)
Track and field: • Sprint (100, 200, and 400 m) • Middle distance (800 and 1500 m) • Hurdles (100, 110, 400 m) • Jumps • Throws	Soccer Rugby American football Basketball Volleyball Handball Baseball Cricket Field hockey	Track and field: • Long distance (5000 and 10,000 m) • 3000 m steeplechase • Walks (20 and 50 km) • Marathon
		Cycling
Tennis		Cross-country skiing
Fencing		Downhill skiing
Gymnastics		Ice hockey
Boxing		Ice skating
Golf		High-altitude sports
Weightlifting		Triathlon
Body building		Pentathlon
Martial arts		Swimming
		Water polo

It is of interest to report that when the risk factors "type of sport" and "atopy" are combined in a logistic regression model, the relative risk of asthma is considerably high: 25-fold in atopic speed and power athletes, 42-fold in atopic long-distance runners and 97-fold in atopic swimmers compared with non-atopic control subjects [70].

No clear classification of EIB endotypes has yet been adopted in clinical practice, and the distinction between T2-high and non-T2-high variants is currently made only by exclusion of a T2-high signature. Therefore, the lack of specific functional and inflammatory biomarkers to classify T2-low EIB subjects currently prevents clear identification of these endotypes and, therefore, an adequate design of clinical research trials. Proper endotyping of EIB would also have extremely relevant clinical translational impact, by significantly contributing to improved disease assessment and management. In fact, although prevention and treatment of EIB is mainly based on the effective use of short- and long-acting β2-agonists, high heterogeneity in individual therapeutic responses and the occurrence of tolerance and side effects have been observed, suggesting the existence of subpopulations requiring an endotype-driven approach to optimize therapy. Moreover, while several targeted therapies are available or under development for a "precision medicine" in T2-high asthma, no tailored strategies are currently available for the non-T2-high forms, in which corticosteroids have shown to be poorly effective. This could be also of relevance in professional athletes in view of the WADA regulations.

4.8 Conclusions

Although regular physical activity is strongly recommended for a proper prevention and management of asthma, evidence has been accumulating that intense and repeated exercise is associated with a higher prevalence of both EIBa and EIBwa. This is particularly true for endurance, winter sports and swimming. Furthermore, in athletes, EIB seems to be only partly reversible, representing exercise itself a causative factor of airway inflammation and symptoms. However, it is reassuring that, when properly diagnosed and optimally treated, athletes with EIB are able to participate on the highest level with their peers [71] with even more chances to succeed and win medals than others in the Olympic Games and other major international competitions [72]. Further research is at last desirable in order to fully address the unmet needs outlined above and to allow the vast population of subjects undertaking physical activity to fully profit from the very beneficial effects of exercise, without incurring health risks or affecting their performance and quality of life.

Key Points

- Exercise-induced bronchoconstriction (EIB) is defined as the transient airway narrowing that occurs as a result of exercise.
- Current guidelines recommend to distinguish EIB with underlying clinical asthma (EIBa) from the occurrence of exercise-induced bronchial obstruction in subjects without other symptoms and signs of asthma (EIBwa).
- The intensity, duration and type of training have been associated with the occurrence of EIB with higher prevalence rates in endurance disciplines, winter sports and swimming.
- When properly managed, EIB does not restrict exercise performance and does not prevent competition at elite level.

References

1. Latimer-Cheung AE, Toll BA, et al. Promoting increased physical activity and reduced inactivity. Lancet. 2013;381(9861):114.
2. Moreira A, Bonini M, Pawankar R, Anderson SD, Carlsen KH, Randolph C, Silvers W, Storms W, Weiler JM, Bonini S. A World Allergy Organization international survey on physical activity as a treatment option for asthma and allergies. World Allergy Organ J. 2014;7(1):34.
3. Moreira A, Delgado L, Haahtela T, Fonseca J, Moreira P, Lopes C, Mota J, Santos P, Rytilä P, Castel-Branco MG. Physical training does not increase allergic inflammation in asthmatic children. Eur Respir J. 2008;32(6):1570–5.
4. Eichenberger PA, Diener SN, Kofmehl R, Spengler CM. Effects of exercise training on airway hyperreactivity in asthma: a systematic review and meta-analysis. Sports Med. 2013;43:1157–70.
5. Del Giacco SR, Garcia-Larsen V. Aerobic exercise training reduces bronchial hyperresponsiveness and serum pro-inflammatory cytokines in patients with asthma. Evid Based Med. 2016;21(2):70.

6. Parsons JP, Hallstrand TS, Mastronarde JG, et al. An official American Thoracic Society clinical practice guideline: exercise-induced bronchoconstriction. Am J Respir Crit Care Med. 2013;187(9):1016.
7. Walsh NP, et al. Position statement. Part one: immune function and exercise. Exerc Immunol Rev. 2011;17:6.
8. Lakier Smith L. Overtraining, excessive exercise, and altered immunity: is this a T helper-1 versus T helper-2 lymphocyte response? Sports Med. 2003;33(5):347.
9. Del Giacco SR, Firinu D, Bjermer L, Carlsen KH. Exercise and asthma: an overview. Eur Clin Respir J. 2015;2:27984.
10. Jones KS, Buston MH, Wharton MJ. The effect of exercise on ventilator function in the child with asthma. Br J Dis Chest. 1962;56:78–86.
11. Bonini M, Palange P. Exercise-induced bronchoconstriction: new evidence in pathogenesis, diagnosis and treatment. Asthma Res Pract. 2015;1:2.
12. Godfrey S. Exercise-induced asthma. In: Clark TJH, Godfrey S, editors. Asthma. London: Chapman and Hall; 1977. p. 57–8.
13. Global Initiative on Asthma. http://ginasthma.org/. Accessed July 2018.
14. Wenzel SE. Asthma: defining the persistent asthma phenotypes. Lancet. 2006;368:804–13.
15. Bonini S. EIB or not EIB? That is the question. Med Sci Sports Exerc. 2008;40(9):1565–6.
16. Weiler JM, Anderson SD, Randolph C, Bonini S, Craig TJ, Pearlman DS, et al. Pathogenesis, prevalence, diagnosis, and management of exercise-induced bronchoconstriction: a practice parameter. Ann Allergy Asthma Immunol. 2010;105:S1–S47.
17. Smoliga JM, Weiss P, Rundell KW. Exercise induced bronchoconstriction in adults: evidence based diagnosis and management. BMJ. 2016;352:h6951.
18. Carlsen KH, Anderson SD, Bjermer L, et al. Exercise-induced asthma, respiratory and allergic disorders in elite athletes: epidemiology, mechanisms and diagnosis: part I of the report from the Joint Task Force of the European Respiratory Society (ERS) and the European Academy of Allergy and Clinical Immunology (EAACI) in cooperation with GA2LEN. Allergy. 2008;63(4):387–403.
19. Randolph C. Exercise-induced bronchospasm in children. Clin Rev Allergy Immunol. 2008;34(2):205–16.
20. Ventura MT, Cannone A, Sinesi D, et al. Sensitization, asthma and allergic disease in young soccer players. Allergy. 2009;64(4):556–9.
21. Bousquet J, Van Cauwenberge P, Khaltaev N, Aria Workshop Group, World Health Organization. Allergic rhinitis and its impact on asthma. J Allergy Clin Immunol. 2001;108(5 Suppl):S147–334.
22. Sandrock CE, Norris A. Infection in severe asthma exacerbations and critical asthma syndrome. Clin Rev Allergy Immunol. 2015;48(1):104–13.
23. Voy RO. The US Olympic committee experience with exercise-induced bronchospasm, 1984. Med Sci Sports Exerc. 1986;18:328–30.
24. Weiler JM, Layton T, Hunt M. Asthma in United States Olympic athletes who participated in the 1996 summer games. J Allergy Clin Immunol. 1998;102:722–6.
25. Katelaris CH, Carrozzi FM, Burke TV, et al. A springtime olympics demands special consideration for allergic athletes. J Allergy Clin Immunol. 2000;106:260–6.
26. Bonini M, Gramiccioni C, Fioretti D, et al. Asthma, allergy and the Olympics: a 12-year survey in elite athletes. Curr Opin Allergy Clin Immunol. 2015;15:184–92.
27. Pignataro FS, Bonini M, Forgione A, et al. Asthma and gender: the female lung. Pharmacol Res. 2017;119:384–90.
28. Romberg K, Tufvesson E, Bjermer L. Sex differences in asthma in swimmers and tennis players. Ann Allergy Asthma Immunol. 2017;118(3):311–7.
29. Norqvist J, Eriksson L, Söderström L, et al. Self-reported physician-diagnosed asthma among Swedish adolescent, adult and former elite endurance athletes. J Asthma. 2015;52(10):1046–53.
30. Settipane RJ, Hagy GW, Settipane GA. Long-term risk factors for developing asthma and allergic rhinitis: a 23-year follow-up study of college students. Allergy Proc. 1994;15:21–5.

31. Braman SS, Barrows AA, De Cotiis BA, et al. Airway hyperresponsiveness in allergic rhinitis: a risk factor for asthma. Chest. 1987;91:671–4.
32. Silvers WS, Poole JA. Exercise-induced rhinitis: a common disorder that adversely affects allergic and non-allergic athletes. Ann Allergy Asthma Immunol. 2006;96(2):334–40.
33. Bonini S, Bonini M, Bousquet J, et al. Rhinitis and asthma in athletes: an ARIA document in collaboration with GA2LEN. Allergy. 2006;61(6):681–92.
34. Helenius IJ, Rytila¨ P, Sarna S, et al. Effect of continuing or finishing high-level sports on airway inflammation, bronchial hyperresponsiveness, and asthma: a 5-year prospective follow-up study of 42 highly trained swimmers. J Allergy Clin Immunol. 2002;109:962–8.
35. Anderson SD, Daviskas E. The mechanism of exercise-induced asthma is …. J Allergy Clin Immunol. 2000;106:453–9.
36. McFadden ER. Hypothesis: exercise-induced asthma as a vascular phenomenon. Lancet. 1990;1:880–3.
37. Couto M, Kurowski M, Moreira A, et al. Mechanisms of exercise-induced bronchoconstriction in athletes: current perspectives and future challenges. Allergy. 2018;73:8–16.
38. Kurowski M, Jurczyk J, Jarzębska M, et al. Serum but not exhaled breath condensate periostin level is increased in competitive athletes. Clin Respir J. 2018;14(1):60–8.
39. Gleeson M, Pyne DB. Respiratory inflammation and infections in high-performance athletes. Immunol Cell Biol. 2016;94(2):124–31.
40. Chimenti L, et al. Bronchial epithelial damage after a half-marathon in non-asthmatic amateur runners. Am J Physiol Lung Cell Mol Physiol. 2010;298(6):L857.
41. Bonini M, et al. Increased nerve growth factor serum levels in top athletes. Clin J Sport Med. 2013;23(3):228.
42. Tufvesson E, Svensson H, Ankerst J, et al. Increase of club cell (Clara) protein (CC16) in plasma and urine after exercise challenge in asthmatics and healthy controls, and correlations to exhaled breath temperature and exhaled nitric oxide. Respir Med. 2013;107(11):1675–81.
43. Bonini M, Usmani OS. The role of the small airways in the pathophysiology of asthma and chronic obstructive pulmonary disease. Ther Adv Respir Dis. 2015;9(6):281–93.
44. Stang J, Stensrud T, Mowinckel P, et al. Parasympathetic activity and bronchial hyperresponsiveness in athletes. Med Sci Sports Exerc. 2016;48(11):2100–7.
45. Bonini M, Braido F, Baiardini I, et al. AQUA: allergy questionnaire for athletes. Development and validation. Med Sci Sports Exerc. 2009;41(5):1034–41.
46. Ansley L, Kippelen P, Dickinson J, Hull JH. Misdiagnosis of exercise-induced bronchoconstriction in professional soccer players. Allergy. 2012;67(3):390–5.
47. Bonini M, Lapucci G, Petrelli G, Todaro A, Pamich T, Rasi G, et al. Predictive value of allergy and pulmonary function tests for the diagnosis of asthma in elite athletes. Allergy. 2007;62(10):1166–70.
48. Kaminsky DA, Irvin CG, Gurka DA, Feldsien DC, Wagner EM, Liu MC, Wenzel SE. Peripheral airways responsiveness to cool, dry air in normal and asthmatic individuals. Am J Respir Crit Care Med. 1995;152(6 Pt 1):1784–90.
49. Crapo RO, Casaburi R, Coates AL, Enright PL, Hankinson JL, Irvin CG, MacIntyre NR, McKay RT, Wanger JS, Anderson SD, Cockcroft DW, Fish JE, Sterk PJ. Guidelines for methacholine and exercise challenge testing-1999. This official statement of the American Thoracic Society was adopted by the ATS Board of directors, July 1999. Am J Respir Crit Care Med. 2000;161(1):309–29.
50. Carlsen KH, Engh G, Mørk M. Exercise-induced bronchoconstriction depends on exercise load. Respir Med. 2000;94(8):750–5.
51. Hull JH, Ansley L, Price OJ, et al. Eucapnic voluntary Hyperpnea: gold standard for diagnosing exercise-induced bronchoconstriction in athletes? Sports Med. 2016;46(8):1083–93.
52. Bonini M, Palange P. Anaphylaxis and sport. Curr Opin Allergy Clin Immunol. 2014;14(4):323–7.
53. Stickland MK, Rowe BH, Spooner CH, Vandermeer B, Dryden DM. Effect of warm-up exercise on exercise-induced bronchoconstriction. Med Sci Sports Exerc. 2012;44(3):383–91.

54. Millqvist E, Bengtsson U, Löwhagen O. Combining a beta2-agonist with a face mask to prevent exercise-induced bronchoconstriction. Allergy. 2000;55(7):672–5.
55. Moreira A, Bonini M, Garcia-Larsen V, Bonini S, Del Giacco SR, Agache I, Fonseca J, Papadopoulos NG, Carlsen KH, Delgado L, Haahtela T. Weight loss interventions in asthma: EAACI evidence-based clinical practice guideline (part I). Allergy. 2013;68(4):425–39.
56. Heffler E, Bonini M, Brussino L, Solidoro P, Guida G, Boita M, Nicolosi G, Bucca C. Vitamin D deficiency and exercise-induced laryngospasm in young competitive rowers. Appl Physiol Nutr Metab. 2016;41(7):735–40.
57. Koh MS, Tee A, Lasserson TJ, Irving LB. Inhaled corticosteroids compared to placebo for prevention of exercise induced bronchoconstriction. Cochrane Database Syst Rev. 2007;3:CD002739.
58. Anderson SD, Caillaud C, Brannan JD. Beta2-agonists and exercise-induced asthma. Clin Rev Allergy Immunol. 2006;31(2–3):163–80.
59. Bonini M, Di Mambro C, Calderon MA, Compalati E, Schünemann H, Durham S, Canonica GW. Beta2-agonists for exercise-induced asthma. Cochrane Database Syst Rev. 2013;10:CD003564.
60. Nelson JA, Strauss L, Skowronski M, Ciufo R, Novak R, McFadden ER Jr. Effect of long-term salmeterol treatment on exercise-induced asthma. N Engl J Med. 1998;339(3):141–6.
61. Salpeter SR, Ormiston TM, Salpeter EE. Meta-analysis: respiratory tolerance to regular beta2-agonist use in patients with asthma. Ann Intern Med. 2004;140(10):802–13.
62. Hancox RJ, Subbarao P, Kamada D, Watson RM, Hargreave FE, Inman MD. Beta2-agonist tolerance and exercise-induced bronchospasm. Am J Respir Crit Care Med. 2002;165(8):1068–70.
63. Salpeter SR, Ormiston TM, Salpeter EE. Cardiovascular effects of beta-agonists in patients with asthma and COPD: a meta-analysis. Chest. 2004;125(6):2309–21.
64. Salpeter SR, Wall AJ, Buckley NS. Long-acting beta-agonists with and without inhaled corticosteroids and catastrophic asthma events. Am J Med. 2010;123(4):322–8. e2
65. Spooner CH, Spooner GR, Rowe BH. Mast-cell stabilising agents to prevent exercise-induced bronchoconstriction. Cochrane Database Syst Rev. 2003;4:CD002307.
66. Pearlman DS, van Adelsberg J, Philip G, Tilles SA, Busse W, Hendeles L, Loeys T, Dass SB, Reiss TF. Onset and duration of protection against exercise-induced bronchoconstriction by a single oral dose of montelukast. Ann Allergy Asthma Immunol. 2006;97(1):98–104.
67. Boaventura LC, Araujo AC, Martinez JB, Vianna EO. Effects of ipratropium on exercise-induced bronchospasm. Int J Sports Med. 2010;31(7):516–20.
68. Bonini M, Scichilone N. Tiotropium in asthma: back to the future of anticholinergic treatment. Clin Mol Allergy. 2017;15:20.
69. Fisk MZ, Steigerwald MD, Smoliga JM, et al. Asthma in swimmers: a review of the current literature. Phys Sportsmed. 2010;38(4):28–34.
70. Helenius IJ, Tikkanen HO, Sarna S, et al. Asthma and increased bronchial responsiveness in elite athletes: atopy and sport event as risk factors. J Allergy Clin Immunol. 1998;101:646–52.
71. Bonini M, Bachert C, Baena-Cagnani CE, et al. ARIA initiative, in collaboration with the WHO collaborating center for asthma, rhinitis. What we should learn from the London Olympics. Curr Opin Allergy Clin Immunol. 2013;13(1):1–3.
72. McKenzie DC, Fitch KD. The asthmatic athlete: inhaled beta-2 agonists, sport performance, and doping. Clin J Sport Med. 2011;21(1):46–50.

Exercise in Chronic Obstructive Pulmonary Disease

5

Pierantonio Laveneziana and Paolo Palange

Abstract

Dyspnoea and exercise limitation are among the most common symptoms experienced by patients with chronic obstructive pulmonary disease (COPD) and are linked to poor perceived health status and increased mortality. Cardiopulmonary exercise testing (CPET) provides a unique opportunity to objectively evaluate the ability of the respiratory system to respond to measured physiological stress across the spectrum of disease severity. In symptomatic mild COPD, the combined abnormalities of increased wasted ventilation leading to increased ventilatory demand and critical erosion of the dynamic inspiratory reserve volume lead to intolerable respiratory discomfort and early exercise limitation. In moderate-to-severe COPD, these major physiological abnormalities that culminate in pronounced demand-capacity imbalance of the respiratory system and dyspnoea become further amplified and are evident at relatively low exercise intensities. In this group, exercise intolerance is often further compounded by the effects of generalized skeletal muscle deconditioning and worsening cardio-circulatory dysfunction. Last but not least, significant dynamic lung hyperinflation on

P. Laveneziana (✉)
Sorbonne Université, INSERM, UMRS1158 Neurophysiologie Respiratoire Expérimentale et Clinique, Paris, France

AP-HP, Groupe Hospitalier Pitié-Salpêtrière Charles Foix, Service des Explorations Fonctionnelles de la Respiration, de l'Exercice et de la Dyspnée du Département "R3S", Paris, France
e-mail: pierantonio.laveneziana@aphp.fr

P. Palange
Department of Public Health and Infectious Diseases, Sapienza University of Rome, Rome, Italy

Lorrilard-Spencer Foundation, Sapienza University of Rome, Rome, Italy
e-mail: paolo.palange@uniroma1.it

A. Cogo et al. (eds.), *Exercise and Sports Pulmonology*,
https://doi.org/10.1007/978-3-030-05258-4_5

exertion has recently been demonstrated in asymptomatic subjects with mild airway obstruction. It goes without saying that identification of specific physiological derangements exposed by CPET facilitates an individualized approach to management in COPD.

5.1 Introduction

Exercise intolerance in chronic obstructive pulmonary disease (COPD) is characteristically multifactorial, and the dominant source is often difficult to determine [1]. The relative contribution of individual sources of limitation varies across the stages of disease severity, being also modulated by comorbidities and chronic inactivity. Clinical enquiry and resting pulmonary function tests are notoriously unreliable in predicting patients' physical impairment on exertion. It is widely accepted that a key contributory factor is intolerable dyspnoea as a result of the combination of high ventilatory demand and abnormal dynamic respiratory mechanics. In this context, the objective assessment of neurosensory, metabolic, cardio-circulatory and respiratory responses to increasing physiological stress (cardiopulmonary exercise testing, CPET) might provide unique informations on the mechanisms and consequences of exertional dyspnoea in these patients. CPET variables (e.g. peak oxygen uptake, $V'O_2$) and minute ventilation-carbon dioxide output (V'_E–$V'CO_2$) relationship, as well as other exercise capacity indices, can also predict earlier mortality in these patients [2–4]. In this chapter, we will briefly discuss the clinical utility of CPET in COPD, and we will then explore the causes of high ventilatory demand and dynamic mechanical impairment in COPD and contrast the responses to CPET in mild and moderate-to-severe COPD. We will review cardiovascular abnormalities during exercise in COPD and current concepts of mechanisms of exertional dyspnoea. Finally, we will touch on sensory and physiological responses to CPET in the emerging area of asymptomatic subjects with airway obstruction.

5.2 Evaluating Exercise Limitation in COPD: Methodological Considerations

A structured approach to CPET interpretation should always consider the evaluation of the following variables reflecting (Fig. 5.1):

- Sensory responses to effort: dyspnoea (Borg) ratings as a function of work rate (WR) and/or V'_E (Fig. 5.2).
- Ventilatory control: V'_E, V'_E–$V'CO_2$ ratio, O_2 saturation by pulse oximetry (SpO_2) and end-tidal partial pressure for CO_2 ($P_{ET}CO_2$) as a function of WR.
- Dynamic respiratory mechanics: change in inspiratory capacity (IC), inspiratory reserve volume (IRV), tidal volume (V_T) and breathing frequency (f) as a function of WR or V'_E and flow-volume loop analysis (Fig. 5.3).
- Metabolic and cardio-circulatory responses: $V'O_2$-WR, $V'CO_2$- $V'O_2$, heart rate (HR) and O_2 pulse as a function of $V'O_2$ [5–12].

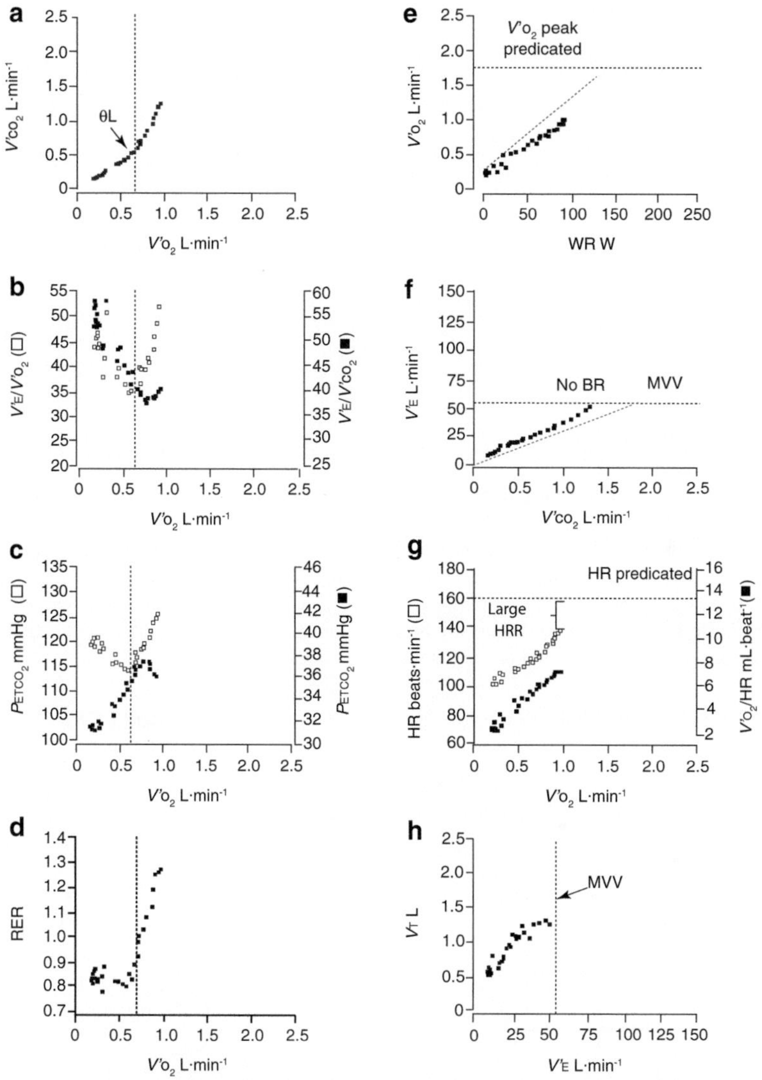

Fig. 5.1 Basic CPET display format recommended by the authors. Plots (**a–d**), in addition to peak $V'O_2$, provide the variables recommended to indirectly estimate the θL, i.e. the $V'O_2$ at which supplemental $V'CO_2$ first becomes evident, in association with hyperventilation relative to O_2 (increased $V'_E/V'O_2$ and $P_{ET}O_2$) but not CO_2 (no increase in $V'_E/V'CO_2$ and no decrease in $P_{ET}CO_2$). Plot (**e**) ($V'O_2$ versus WR) provides an indication of exercise and O_2 transport efficiency and also peak $V'O_2$. Plots (**f**) (V'_E versus $V'CO_2$) and (**h**) (V_T versus V'_E) characterize aspects of the ventilatory response. Finally, plot (**g**) is informative with respect to the characteristics of the haemodynamic response. Abbreviations: *$V'O_2$* oxygen uptake, *$V'CO_2$* carbon dioxide output, *$V'_E/V'O_2$* ventilatory equivalent for O_2, *$V'_E/V'CO_2$* ventilatory equivalent for CO_2, *$P_{ET}O_2$* end tidal partial pressure of O_2, *$P_{ET}CO_2$* end tidal partial pressure of CO_2, *RER* respiratory exchange ratio, *θL* lactate threshold, *WR* work rate, *V'_E* minute ventilation, *HR* heart rate, *HRR* heart rate reserve, *V_T* tidal volume, *MVV* maximal voluntary ventilation, *BR* breathing reserve. $V'O_2$-WR slope and V'_E–$V'CO_2$ slope: red dotted line, best fit to linear region

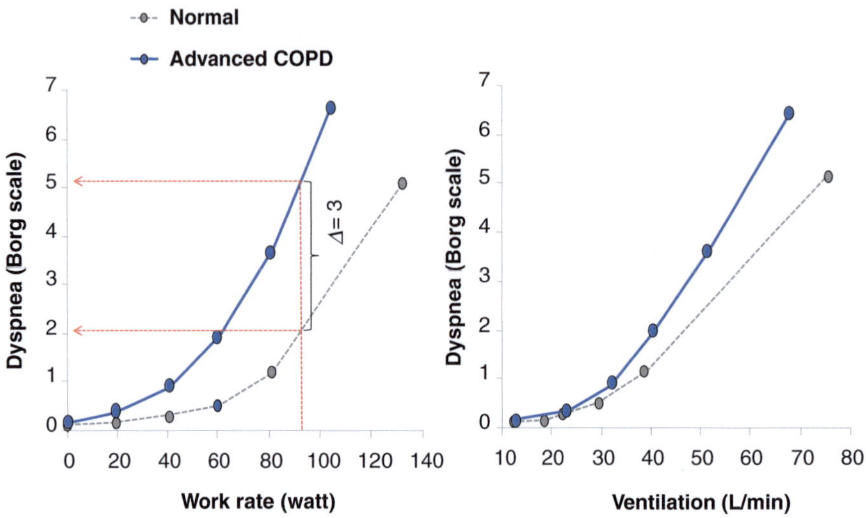

Fig. 5.2 Dyspnoea intensity responses during symptom-limited incremental cycle CPET in an advanced COPD patient (solid line) and an age- and sex-matched healthy control subject (dashed line). Exertional dyspnoea intensity (Borg score) is shown in response to increasing WR (**a**) and V'_E (**b**). Differences in dyspnoea intensity between the COPD patient and the healthy subject are shown at iso-WR. Δ difference. Further details are provided in the text

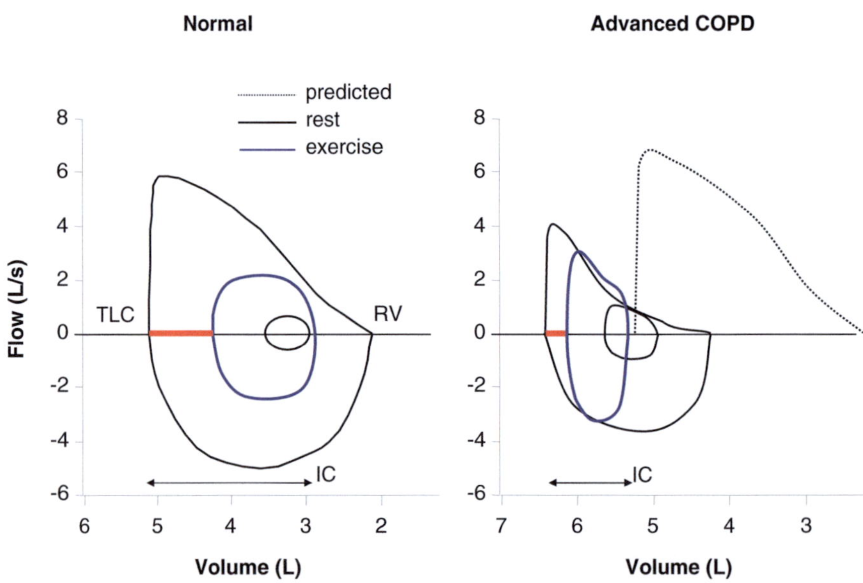

Fig. 5.3 Flow-volume loops showing the effects of exercise in (**a**) a healthy control subject and (**b**) a patient with advanced COPD. The outer black loop shows the maximal limits of flow and volume (the control loop is shown as a dashed profile in **b**); the inner black loop is the resting profile; the blue loop is the maximal exercise profile. The red line is the IRV. Healthy subjects are able to increase both their V_T and inspiratory and expiratory flows during exercise. In COPD, expiratory flow may already be maximal at rest. In order to increase expiratory flow during exercise, these patients must hyperinflate. Further details are provided in the text

5.3 Measuring Perceptual Responses to Exercise

The intensity of leg and breathing discomfort (dyspnoea) can be measured by validated scales such as the Borg scale—a category scale with ratio properties—and visual analogue scales, both expressed as a function of increasing WR, $V'O_2$ or V'_E [13, 14]. These scales show good test-retest reproducibility and responsiveness to therapeutic interventions [15]. The dominant exercise-limiting symptom (i.e., dyspnoea, leg discomfort, a combination of both or other) can be determined at the end of the test by simply asking "why did you stop exercising?: was it because of your breathing, your legs, a mixture of both, or for some other reason?".

5.4 Assessment of Respiratory System Reserve During Exercise

A low breathing reserve, usually expressed as the difference between maximal voluntary ventilation (MVV) [or maximal ventilatory capacity] and peak V'_E (MVV-peak V'_E < 15 L/min), is used as a traditional indicator of ventilatory limitation in COPD and in other respiratory diseases [16], although it has some limitations:

- Respiratory sensation, operating lung volumes and respiratory muscle recruitment patterns are quite different during short bursts of voluntary hyperventilation compared with physiological conditions at peak V'_E during CPET.
- The proximate limitation of exercise performance in chronic lung diseases is very often intolerable symptoms such as dyspnoea, leg discomfort or both. Thus, the low peak V'_E and apparent preserved breathing reserve may simply reflect early symptom limitation well before the physiological limits of the respiratory and cardio-circulatory are reached [17].

Accordingly, the assessment of breathing reserve during standardized submaximal WR compared with age- and sex-matched controls can provide more important information than simple peak comparisons. Moreover, there is now broad acknowledgement that measurements of exertional symptoms and non-invasive dynamic respiratory mechanics are an integral component of the evaluation of the prevailing ventilatory constraints, during CPET [11, 15, 18, 19].

5.5 Breathing Pattern and Operating Lung Volumes During Exercise

Compared to health, breathing pattern in COPD is more shallow and fast. This is best seen when V_T and f are expressed as a function of increasing V'_E. V_T expands early in exercise until it reaches approximately 70% of the resting IC after which an inflection or plateau is seen. The plateau occurs when end-inspiratory lung volume (EILV) reaches a minimal IRV about 0.5–1.0 L below total lung capacity (TLC). In COPD, this plateau represents a true limitation to further V_T expansion since, in contrast to the situation in health, added dead space (equivalent to CO_2 re-breathing) has no effect on volume expansion [20]. Interestingly, when the increase in chemo-stimulation occurs in the setting of a fixed maximal V_T, dyspnoea quickly increases [20, 21]. At this point the only option to respond to the increasing drive of exercise is an increased breathing frequency. Serial IC measurements throughout exercise allow us to track the change in end-expiratory lung volume (EELV) on the reasonable assumption that TLC remains unaltered [11]. The average change in IC above the baseline value has been reported as 0.3–0.6 L and depends on the resting IC. Thus, the lower the resting IC (the greater the resting lung hyperinflation), thc less the decrease in this variable with increasing V'_E during exercise [15].

5.6 Flow-Volume Loop Analysis During Exercise

A comparison of tidal and maximal flow-volume loops at rest and during exercise provides valuable "non-invasive" assessments of dynamic respiratory mechanics during exercise and breathing pattern [12]. Evaluation of reserves for respired flow and volume generation at any point during exercise is particularly useful but requires accurate placement of EELV on the volume axis by serial IC measurements (Fig. 5.3) [11]. Volume correction for actual absolute TLC in each individual is desirable as is accounting for possible configurational changes of the maximal envelope due to natural intrinsic bronchodilation or, in some instances, acute bronchoconstriction during exercise. Possible variation in airway smooth muscle tone during exercise can be detected by comparing maximal flow-volume curves at rest prior to exercise and immediately postexercise. In COPD, the main abnormalities are:

- Expiratory flow limitation as assessed by the degree of progressive overlap of tidal expiratory flows on the maximal expired flow envelope.
- Dynamic lung hyperinflation.
- Shift of tidal volume loops to the left as IC decreases.
- Critical mechanical constraints on V_T expansion as manifest by a low IRV at the limits of tolerance (Fig. 5.3).

While flow-volume loop analysis provides mainly qualitative evaluation (pattern recognition), quantitative assessments of dynamic breathing reserves during exercise are increasingly provided in exercise test read-outs, at least in research settings [12].

5.7 Increased Ventilatory Demand During Exercise in COPD

The ventilatory response to exercise is geared to improve pulmonary gas exchange while minimizing the work of breathing and attendant respiratory sensations. In addition to the predations of ageing, exercise ventilation in COPD is perturbed by smoking-related inflammatory injury of the airways, lung parenchyma and microvasculature. Each of these structural abnormalities conspires against the normal coupling of V'_E to metabolic demand ($V'CO_2$) as:

$$V'_E / V'CO_2 = 1/\left[PaCO_2 \times \left(1 - V_D / V_T\right)\right] \quad (5.1)$$

where $PaCO_2$ is the partial pressure of CO_2 in the arterial blood and V_D/V_T is dead space-to-tidal volume ratio. Thus, the higher the $V'_E/V'CO_2$ (i.e. ventilation is less "efficient"), the higher the fraction of the breath wasted in the V_D [16]. When concomitant cardiovascular disease is present, $PaCO_2$ may be regulated to a lower level (i.e. low CO_2 set-point), reflecting adaptations of the central regulatory controller [22]. In symptomatic smokers with largely preserved forced expiratory volume in 1 s (FEV_1), $V'_E/V'CO_2$ is frequently increased [6–9, 20, 23, 24] due to an enlarged V_D rather than a small V_T or a low $PaCO_2$ set-point [7]. In line with these premises, adding external V_D further increases $V'_E/V'CO_2$ in these patients [20]. Low $P_{ET}CO_2$ is also commonly seen because the expired CO_2 is diluted by the air that comes from areas of high ventilation/perfusion (V'_A/Q') ratio due to poor pulmonary perfusion [7, 25]. The excessive ventilatory response decreases patients' mechanical reserves, thereby contributing to exertional dyspnoea and poor exercise tolerance [6–9, 20]. Ventilation responses to incremental exercise may also be increased in COPD by increased central chemo-stimulation due to early metabolic acidosis [26]. This usually occurs as a result of skeletal muscle (reduced oxidative capacity) and cardiac deconditioning—an inevitable consequence of prolonged physical inactivity due to dyspnoea [27]. Critical hypoxaemia (partial pressure of arterial O_2 (PaO_2) < 60 mmHg) can also stimulate ventilation via peripheral chemoreceptors as a result of decreased mixed venous O_2 in systemic blood perfusing alveolar units with low V'_A/Q' ratios or right-to-left shunting. Type I and II ergoreceptor stimulation in response to an acidic milieu in the peripheral muscles in conjunction with sympathetic over-activation are additional sources of ventilatory stimulation in COPD [28]. Finally, as already mentioned, in patients with COPD and concomitant congestive heart failure, a low regulated CO_2 set-point in the setting of a fixed high physiological dead space (ventilatory inefficiency) amplifies the ventilatory response to exercise [22].

5.8 Abnormal Lung Mechanics During Exercise in COPD

The key mechanical abnormalities which are present at rest and worsened by exercise in COPD include [10, 29–31]:

- High elastic loading which is amplified by increased inspiratory threshold loading due to the effect of intrinsic positive end-expiratory pressure.
- Decreased dynamic lung compliance.
- Increased resistive loading of the respiratory muscles.

All of these abnormalities are mechanistically linked to increased respiratory motor drive and inspiratory muscle effort. In the presence of expiratory flow limitation and faster f (leading to shorter expiratory time), there is a temporary increase in EELV above the resting value, i.e. acute-on-chronic lung hyperinflation. As TLC does not change appreciably with exercise, IC decreases and VT lies closer to the upper portion of pressure-volume relation of the respiratory system, where compliance is decreased and the inspiratory muscles are functionally weakened. In fact, EILV encroaches on TLC leading to a critically reduced IRV and dynamic mechanical constraints [10, 11, 20, 32]. This explains why V_T cannot further increase and the consequent relative tachypnoea in COPD compared with healthy controls [10, 20]. Hyperinflation also promotes functional inspiratory muscle weakness as the diaphragm loses efficiency when overstretched and higher f imposes an excessively fast rate of shortening [33].

Expiratory muscle activity is relatively increased in COPD but fails to prevent dynamic hyperinflation [34, 35]. Excessive expiratory muscle activity will cause increased dynamic compression of the airways on expiration without increasing tidal expiratory flow rates or reducing EELV and may have deleterious hemodynamic effects (e.g. reduced venous return), which further compromise exercise performance [29, 36]. It remains unclear whether some patients with more advanced COPD do present with respiratory muscle fatigue at the end of a symptom-limited exercise test. Dynamic functional weakness as a consequence of gas trapping and increased velocity on contraction, however, is more likely [37]. Resting inspiratory muscle weakness under "static" conditions due to systemic myopathy, cachexia, malnutrition and repeated steroids usage has been reported in a minority of subjects [38, 39]. Acute-on-chronic decrement in respiratory muscle strength might occur with exercise progression in these patients [40].

5.9 Cardiovascular Responses to Exercise in COPD

There is growing evidence that patients in early stages of COPD present with cardio-circulatory abnormalities [41, 42]. This is not surprising considering the common risk factor of smoking and low-grade inflammation. Emphysema in mild COPD is more closely associated with systemic vascular dysfunction than airway disease [43], suggesting that inflammatory-immunological mechanisms involved in lung elastolysis might act remotely leading to early loss of vascular elasticity in large arteries and impairment in endothelial function in peripheral arteries [44–46]. Considering the significant relationship between emphysema burden and exertional dyspnoea in mild COPD [47], it is also possible that patients with worse emphysema are particularly inactive leading to poorer systemic vascular function. Resting heart rate (HR) tends to be higher in patients with mild COPD, an abnormality which persists during exercise [7, 8, 20]. Increase in HR, however, is not severe enough to impair $V'O_2$/HR ratio (oxygen pulse) or decrease HR reserve (predicted maximal HR minus peak HR) in patients with mild-moderate COPD and smokers with relatively preserved pulmonary function [7, 48]. Impairment (or destruction/

obliteration) of pulmonary microvasculature might develop in mild COPD [49, 50]; in fact, reduced pulmonary blood flow on exercise (at a given $V'O_2$) has been recently demonstrated in patients with mild COPD [47]. Although one study reported the presence of pulmonary arterial hypertension (PAH) in up to 17% of patients with mild COPD, few patients present with clinically significant PAH at rest [51]. Those with resting PAH, however, present with likelihood of developing PAH during exercise [52], a finding that predicts future resting PAH [53].

In more severe COPD, there is a long-standing concept that high mean intrathoracic pressures secondary to severe gas trapping and lung hyperinflation could impair venous return and right ventricular preload on exercise [54]. This would be particularly worsened if the patient chooses to recruit the abdominal expiratory muscles leading to lower vena cava compression [54]. Right ventricular function could be further impaired as a consequence of increased pulmonary vascular resistance (PVR), i.e. high right ventricular afterload. High PVR in moderate-to-severe COPD might be related to [55–59]:

- The negative mechanical effects of breathing at lung volumes close to TLC.
- Vasoconstriction provoked by low intra-alveolar PO_2 tensions.
- Emphysematous destruction of the vascular bed.
- Mechanical compression of juxta-alveolar vessels due to scattered pockets of dynamic hyperinflation during exercise.

In fact, some studies have shown impaired right ventricular ejection fraction despite increased right ventricular end-diastolic pressure [58–61]. Dynamic impairment in left ventricular function during exercise might be seen in advanced COPD due to the large intrathoracic pressure swings required to overcome increased elastic and resistive loads. This is likely to be particularly deleterious in the presence of cardiac comorbidity, e.g. chronic heart failure [62–64]. In end-stage COPD, severe lung hyperinflation may mechanically impair cardiac output [65, 66]. Left ventricular diastolic function may be impaired due to the negative consequences of ventricular interdependence [60, 67]. There is a paucity of haemodynamic data obtained in invasive studies in COPD: the prevailing view is that cardiac output tends to increase within the expected range (relative to $V'O_2$) despite the increased PVR [63]. Nevertheless, peak cardiac output (and $V'O_2$) reaches lower values compared to healthy subjects, likely due to early exercise termination induced by limiting dyspnoea [56, 68].

5.10 Exercise Pathophysiology in Mild COPD

Symptomatic smokers with only modest physiological abnormalities at rest might present with activity-related dyspnoea leading to chronic inactivity restriction, reduced health-related quality of life and poorer prognosis [69–71]. In this context, CPET has a major role in exposing the heterogeneous nature of the physiological impairment in smokers with minor airway obstruction [6–9, 20, 72]. A number of physiological abnormalities at rest have the potential to impact on exercise responses in patients with mild COPD: increased peripheral airways resistance [73, 74]; heterogeneous distribution of alveolar ventilation [75]; EFL, pulmonary gas trapping and reduced IC [6–9, 20]; increased alveolar-to-arterial O_2 tension gradient (A-aPO_2) [7, 76, 77]; and low lung transfer factor for carbon monoxide and transfer coefficient [78–80].

During incremental CPET, the following abnormalities can be found [7, 20]: increased $V'_E/V'CO_2$ nadir and steeper V'_E–$V'CO_2$ slope compared with healthy controls reflecting heightened chemo-stimulation due to high physiological dead space, and dynamic hyperinflation secondary to expiratory flow limitation, and high ventilatory demand. The combination of reduced IC and high inspiratory neural drive results in early attainment of critical mechanical constraints and higher exertional dyspnoea ratings compared with age-matched healthy controls [6–9, 20].

5.11 Exercise Pathophysiology in Moderate-to-Severe COPD

The above-mentioned pattern of abnormalities is also found in more advanced COPD. However, derangements of pulmonary gas exchange and dynamic respiratory mechanics are seen at significantly lower V'_E and WR [10]. The higher ventilatory requirements relative to metabolic demands are mechanistically linked to increased afferent stimulation of central and peripheral chemoreceptors due to:

- Alveolar ventilation (V'_A)/lung capillary perfusion ($Q'c$) mismatching characterized by high V'_A/Q' lung units and increased "wasted" ventilation in the physiological dead space [23, 81]
- Hypoxaemia due to increased perfusion of low $V'_A/Q'c$ lung units with poorly oxygenated mixed venous blood [82, 83]
- Higher lactacidotic drive due to deconditioning or impaired central haemodynamic [84, 85]

Critical mechanical constraints and inspiratory muscle dysfunction, particularly in the context of high V_D and blunted V_T expansion [22, 86], may lead to alveolar hypoventilation and CO_2 retention in end-stage COPD.

It should be noted that a commonly used metric of ventilatory inefficiency (V'_E–$V'CO_2$ slope) decreases from mild to (very) severe COPD [24]. This paradoxical finding is a consequence of mechanical impediment of the ventilatory response induced by critical inspiratory constraints in more severe COPD [10, 23, 32]. The resulting hypercapnia increases the prevailing levels at which CO_2 is centrally regulated which further contributes to lessen V'_E for a given $V'CO_2$ [86, 87]. These considerations explain why the V'_E–$V'CO_2$ relationship may underestimate the expected increase in physiological dead space in COPD patients with severe mechanical constraints during exercise. Similar considerations apply to some COPD patients with coexistent heart failure who present with particularly high $V'_E/V'CO_2$ on exercise [88]. This abnormality is more closely related to a lower $PaCO_2$ set-point rather than a high physiological dead space [22]. Moreover, low $PaCO_2$ is also seen in COPD-heart failure patients with exercise oscillatory ventilation; of note, those oscillations are blunted or disappear when V_T expansion is limited by critical mechanical constraints. As expected, combination of high chemo-stimulation and impaired lung mechanics prompt severe dyspnoea and particularly poor exercise tolerance in these patients [64].

5.12 Exertional Dyspnoea in COPD

In the majority of patients, intolerable breathlessness is the primary locus of symptom limitation, especially in more advanced disease. However, leg discomfort alone or in combination with dyspnoea is also commonly reported at end-exercise and points to coexistent skeletal muscle weakness and deconditioning or even peripheral vascular disease or abnormal cardiopulmonary interactions [89]. Despite those considerations, CPET is more frequently interrupted by the patients due to early attainment of critical mechanical constraints leading to intolerable respiratory discomfort [32, 90, 91]. As mentioned, low resting IC which did not increase with exercise, or progressive decrease in IC (secondary to dynamic gas trapping), reduces the operating limits for V_T expansion leading to early attainment of a critically low IRV during progressive exercise [32]. This is where the disparity between increasing inspiratory neural drive and the muscular/mechanical response of the respiratory system abruptly widens [i.e. neuromechanical dissociation] and marks the threshold beyond which dyspnoea intensity rises sharply to intolerable levels [32, 92]. Improvement in dynamic lung mechanics (resting and exercise IC) with inhaled bronchodilator has been found of foremost relevance to restore neuromechanical coupling and to delay mechanical limitation [92, 93]. These results are consistent with the notion that activity-related dyspnoea and exercise intolerance are closely related to consequences of gas trapping and lung hyperinflation (i.e. higher inspiratory neural drive and neuromechanical dissociation). In fact, pharmacological lung deflation induced by these medications has been instrumental to delay the dyspnoea threshold and prolongs exercise endurance. Furthermore, strategies aimed at reducing inspiratory neural drive (e.g. supplemental oxygen [94–96], opiates [97–101] and exercise training [102, 103]) have been found useful to lessen dyspnoea and increase exercise tolerance.

5.13 Asymptomatic Subjects with Airway Obstruction

In at-risk subjects who complain of symptoms, early detection of COPD has been proven to be effective and is now recommended. However, even in the absence of COPD-related symptoms, subjects with persistent airway obstruction defined by a post-bronchodilator FEV_1/FVC ratio below the statistically based, age-specific lower limit of normal (LLN) are at risk of premature death and/or of development of respiratory symptoms, suggesting that their identification is pertinent. It has recently been demonstrated that subjects with mild airway obstruction, defined as post-bronchodilator $FEV_1/FVC < LLN$ together with normal FEV_1, who do not complain of any COPD symptom had, however, greater reduction in IC and half of them had significant dynamic lung hyperinflation, higher dyspnoea intensity during incremental exercise and poorer peak exercise capacity than matched healthy subjects with normal spirometry [104].

5.14 Conclusions

The mechanisms underlying breathlessness and exercise intolerance in COPD can be objectively determined through an integrated analysis of mechanical-ventilatory responses (i.e. ventilatory efficiency, breathing pattern, operating lung volumes and tidal flow-volume loops) in addition to serial measurements of dyspnoea intensity. Even in symptomatic mild COPD, the combined abnormalities of decreased ventilatory efficiency—a surrogate for increased wasted ventilation—which increases ventilatory demand and critical erosion of the dynamic inspiratory reserve volume due, in part, to the effects of dynamic lung hyperinflation lead to intolerable respiratory discomfort and early exercise limitation compared with healthy age- and sex-matched controls. In moderate-to-severe COPD, these major combined physiological abnormalities become further amplified and occur at relatively low exercise intensities. In this group exercise intolerance is further compounded by the effects of generalized skeletal muscle deconditioning and cardio-circulatory abnormalities. The corollary is that combined interventions that attenuate high ventilatory demand and improve dynamic respiratory mechanics will reliably ameliorate exertional dyspnoea and delay the onset of exercise limitation. Finally, a significant dynamic lung hyperinflation on exertion, greater dyspnoea intensity and poorer peak exercise capacity, has recently been demonstrated in asymptomatic subjects with mild airway obstruction compared with matched healthy subjects with normal spirometry during incremental exercise. CPET is a well-tailored approach in the identification of specific physiological derangements in all range of COPD.

Key Points

- Dyspnoea and exercise limitation are among the most common symptoms experienced by patients with chronic obstructive pulmonary disease (COPD) and are linked to poor perceived health status and increased mortality.
- Cardiopulmonary exercise testing (CPET) provides an objective evaluation of the ability of the respiratory system to respond to measured physiological stress across the spectrum of disease severity.
- In symptomatic mild COPD, the combination of increased wasted ventilation leading to increased ventilatory demand and reduction of the dynamic inspiratory reserve volume lead to intolerable respiratory discomfort and early exercise limitation.

- In moderate-to-severe COPD, these major physiological abnormalities that culminate in pronounced demand-capacity imbalance of the respiratory system and dyspnoea become further amplified and are evident at relatively low exercise intensities. In this group, exercise intolerance is often further compounded by the effects of skeletal muscle deconditioning and worsening cardio-circulatory dysfunction.
- Significant dynamic lung hyperinflation on exertion, greater dyspnoea intensity and poorer peak exercise capacity, have recently been demonstrated during incremental exercise in asymptomatic subjects with mild airway obstruction compared with matched healthy subjects with normal spirometry.
- Identification of specific physiological derangements exposed by CPET facilitates an individualized approach to management in COPD.

References

1. O'Donnell DE, Laveneziana P, Neder JA. Lung function. In: Kolb M, Vogelmeier CF, editors. Outcomes in clinical trials, vol. 62. European Respiratory Monograph; 2013. p. 127.
2. Waschki B, Kirsten A, Holz O, et al. Physical activity is the strongest predictor of all-cause mortality in patient with COPD: a prospective cohort study. Chest. 2011;140(2):331–42.
3. Oga T, Nishimura K, Tsukino M, Sato S, Hajiro T. Analysis of factors related to mortality in chronic obstructive pulmonary disease: role of exercise capacity and health status. Am J Respir Crit Care Med. 2003;167:544–9.
4. Pinto-Plata VM, Cote C, Cabral H, et al. The 6-min walk distance: change over time and value as a predictor of survival in severe COPD. Eur Respir J. 2004;23:28–33.
5. Weisman IM, Zeballos RJ. An integrated approach to the interpretation of cardiopulmonary exercise testing. Clin Chest Med. 1994;15(2):421–45.
6. Ofir D, Laveneziana P, Webb KA, et al. Mechanisms of dyspnea during cycle exercise in symptomatic patients with GOLD stage I chronic obstructive pulmonary disease. Am J Respir Crit Care Med. 2008;177(6):622–9.
7. Elbehairy AF, Ciavaglia CE, Webb KA, et al., On behalf of the Canadian Respiratory Research Network. Pulmonary gas exchange abnormalities in mild chronic obstructive pulmonary disease. Implications for dyspnea and exercise intolerance. Am J Respir Crit Care Med. 2015;191(12):1384–94.
8. Elbehairy AF, Raghavan N, Cheng S, et al., On behalf of the Canadian respiratory research network. Physiologic characterization of the chronic bronchitis phenotype in GOLD grade 1B COPD. Chest. 2015;147(5):1235–45.
9. Guenette JA, Chin RC, Cheng S, et al. Mechanisms of exercise intolerance in global initiative for chronic obstructive lung disease grade 1 COPD. Eur Respir J. 2014;44(5):1177–87.
10. Faisal A, Alghamdi BJ, Ciavaglia CE, et al. Common mechanisms of dyspnea in chronic interstitial and obstructive lung disorders. Am J Respir Crit Care Med. 2016;193(3):299–309.
11. Guenette JA, Chin RC, Cory JM, et al. Inspiratory capacity during exercise: measurement, analysis, and interpretation. Pulm Med. 2013;2013:956081.
12. Johnson BD, Weisman IM, Zeballos RJ, Beck KC. Emerging concepts in the evaluation of ventilatory limitation during exercise: the exercise tidal flow-volume loop. Chest. 1999;4116(2):488–503.

13. Borg GA. Psychophysical bases of perceived exertion. Med Sci Sports Exerc. 1982;14:377–81.
14. Gift AG. Validation of a vertical visual analogue scale as a measure of clinical dyspnea. Rehabil Nurs. 1989;14(6):323–5.
15. O'Donnell DE, Travers J, Webb KA, et al. Reliability of ventilatory parameters during cycle ergometry in multicentre trials in COPD. Eur Respir J. 2009;34(4):866–74.
16. Wasserman K, Hansen J, Sue DY, et al. Principles of exercise testing and interpretation. 3rd ed. Baltimore, MD: Lippincott, Williams and Wilkins; 1999.
17. Dempsey JA. Limits to ventilation (for sure!) and exercise (maybe?) in mild chronic obstructive pulmonary disease. Am J Respir Crit Care Med. 2013;187(12):1282–3.
18. Puente-Maestu L, Palange P, Casaburi R, et al. Use of exercise testing in the evaluation of interventional efficacy: an official ERS statement. Eur Respir J. 2016;47(2):429–60.
19. ERS Task Force, Palange P, Ward SA, Carlsen K-H, et al. Recommendations on the use of exercise testing in clinical practice. Eur Respir J. 2007;29:185–209.
20. Chin RC, Guenette JA, Cheng S, et al. Does the respiratory system limit exercise in mild chronic obstructive pulmonary disease? Am J Respir Crit Care Med. 2013;187(12):1315–23.
21. O'Donnell DE, Hong HH, Webb KA. Respiratory sensation during chest wall restriction and dead space loading in exercising men. J Appl Physiol (1985). 2000;88(5):1859–69.
22. Rocha A, Arbex FF, Sperandio PA, et al. Excess ventilation in chronic obstructive pulmonary disease-heart failure overlap. Implications for dyspnea and exercise intolerance. Am J Respir Crit Care Med. 2017;196(10):1264–74.
23. O'Donnell DE, Maltais F, Porszasz J, et al., 205.440 Investigators. The continuum of physiological impairment during treadmill walking in patients with mild-to-moderate COPD: patient characterization phase of a randomized clinical trial. PLoS One. 2014;9(5):e96574.
24. Neder JA, Arbex FF, Alencar MC, et al. Exercise ventilatory inefficiency in mild to end-stage COPD. Eur Respir J. 2015;45(2):377–87.
25. Hansen JE, Ulubay G, Chow BF, et al. Mixed-expired and end-tidal CO_2 distinguish between ventilation and perfusion defects during exercise testing in patients with lung and heart diseases. Chest. 2007;132(3):977–83.
26. Casaburi R, Patessio A, Ioli F, et al. Reductions in exercise lactic acidosis and ventilation as a result of exercise training in patients with obstructive lung disease. Am Rev Respir Dis. 1991;143:9–18.
27. Casaburi R. Limitation to exercise tolerance in chronic obstructive pulmonary disease: look to the muscles of ambulation. Am J Respir Crit Care Med. 2003;168(4):409–10.
28. Maltais F, Decramer M, Casaburi R, et al., ATS/ERS ad hoc committee on limb muscle dysfunction in COPD. An official American Thoracic Society/European Respiratory Society statement: update on limb muscle dysfunction in chronic obstructive pulmonary disease. Am J Respir Crit Care Med. 2014;189(9):e15–62.
29. Potter WA, Olafsson S, Hyatt RE. Ventilatory mechanics and expiratory flow limitation during exercise in patients with obstructive lung disease. J Clin Invest. 1971;50:910–9.
30. Dodd DS, Brancatisano T, Engel LA. Chest wall mechanics during exercise in patients with severe chronic air-flow obstruction. Am Rev Respir Dis. 1984;129(1):33–8.
31. Jolley CJ, Luo YM, Steier J, et al. Neural respiratory drive and breathlessness in COPD. Eur Respir J. 2015;45:355–64.
32. O'Donnell DE, Guenette JA, Maltais F, Webb KA. Decline of resting inspiratory capacity in COPD: the impact on breathing pattern, dyspnea, and ventilatory capacity during exercise. Chest. 2012;141(3):753–62.
33. Leblanc P, Summers E, Inman MD, et al. Inspiratory muscles during exercise: a problem of supply and demand. J Appl Physiol (1985). 1988;64:2482–9.
34. Ciavaglia CE, Guenette JA, Langer D, et al. Differences in respiratory muscle activity during cycling and walking do not influence dyspnea perception in obese patients with COPD. J Appl Physiol (1985). 2014;117:1292–301.
35. Laveneziana P, Webb KA, Wadell K, et al. Does expiratory muscle activity influence dynamic hyperinflation and exertional dyspnea in COPD? Respir Physiol Neurobiol. 2014;199:24–33.
36. Kyroussis D, Polkey MI, Hamnegard CH, et al. Respiratory muscle activity in patients with COPD walking to exhaustion with and without pressure support. Eur Respir J. 2000;15:649–55.

37. Mador MJ, Kufel TJ, Pineda LA, Sharma GK. Diaphragmatic fatigue and high-intensity exercise in patients with chronic obstructive pulmonary disease. Am J Respir Crit Care Med. 2000;161:118–23.
38. Gosselink R, Troosters T, Decramer M. Peripheral muscle weakness contributes to exercise limitation in COPD. Am J Respir Crit Care Med. 1996;153:976–80.
39. Charususin N, Gosselink R, McConnell A, et al. Inspiratory muscle training improves breathing pattern during exercise in COPD patients. Eur Respir J. 2016;47:1261–4.
40. Rodrigues A, da Silva M, Berton D, et al. Maximal inspiratory pressure: does the choice of reference values actually matter? Chest. 2017;152(1):32–9.
41. Sin DD, Wu L, Man SF. The relationship between reduced lung function and cardiovascular mortality: a population-based study and a systematic review of the literature. Chest. 2005;127:1952–9.
42. Malerba M, Ragnoli B, Salameh M, et al. Sub-clinical left ventricular diastolic dysfunction in early stage of chronic obstructive pulmonary disease. J Biol Regul Homeost Agents. 2011;25(3):443–51.
43. Zelt JT, Jones JH, Daniel M, Hirai DM, et al. Systemic vascular dysfunction is associated with emphysema burden in mild COPD. Respir Med. 2018;136:29. https://doi.org/10.1016/j.rmed.2018.01.007.
44. Muñoz-Esquerre M, López-Sánchez M, Escobar I, et al. Systemic and pulmonary vascular remodelling in chronic obstructive pulmonary disease. PLoS One. 2016;11(4):e0152987.
45. Eickhoff P, Valipour A, Kiss D, et al. Determinants of systemic vascular function in patients with stable chronic obstructive pulmonary disease. Am J Respir Crit Care Med. 2008;178(12):1211–8.
46. Clarenbach CF, Senn O, Sievi NA, et al. Determinants of endothelial function in patients with COPD. Eur Respir J. 2013;42(5):1194–204.
47. Jones JH, Zelt JT, Hirai DM, et al. Emphysema on thoracic CT and exercise ventilatory inefficiency in mild-to-moderate COPD. COPD. 2017;14(2):210–8.
48. Wang HY, Xu QF, Xiao Y, et al. Cardiac response and N-terminal-pro-brain natriuretic peptide kinetics during exercise in patients with COPD. Respir Care. 2011;56(6):796–9.
49. Grau M, Barr RG, Lima JA, et al. Percent emphysema and right ventricular structure and function: the MESA lung and MESARV studies. Chest. 2013;144(1):136–44.
50. Barr RG, Bluemke DA, Ahmed FS, et al. Percent emphysema, airflow obstruction, and impaired left ventricular filling. N Engl J Med. 2010;362(3):217–27.
51. Gupta NK, Agrawal RK, Srivastav AB, Ved ML. Echocardiographic evaluation of heart in chronic obstructive pulmonary disease patient and its co-relation with the severity of disease. Lung India. 2011;28(2):105–9.
52. Portillo K, Torralba Y, Blanco I, et al. Pulmonary hemodynamic profile in chronic obstructive pulmonary disease. Int J Chron Obstruct Pulmon Dis. 2015;10:1313–20.
53. Kessler R, Faller M, Weitzenblum E, et al. Natural history of pulmonary hypertension in a series of 131 patients with chronic obstructive lung disease. Am J Respir Crit Care Med. 2001;164:219–24.
54. Aliverti A, Macklem PT. How and why exercise is impaired in COPD. Respiration. 2001;68:229–39.
55. Light RW, Mintz HM, Linden GS, Brown SE. Hemodynamics of patients with severe chronic obstructive pulmonary disease during progressive upright exercise. Am Rev Respir Dis. 1984;130:391–5.
56. Mahler DA, Brent BN, Loke J, et al. Right ventricular performance and central circulatory hemodynamics during upright exercise in patients with chronic obstructive pulmonary disease. Am Rev Respir Dis. 1984;130:722–9.
57. Oswald-Mammosser M, Apprill M, Bachez P, et al. Pulmonary hemodynamics in chronic obstructive pulmonary disease of the emphysematous type. Respiration. 1991;58:304–10.
58. Magee F, Wright JL, Wiggs BR, et al. Pulmonary vascular structure and function in chronic obstructive pulmonary disease. Thorax. 1988;43:183–9.
59. Agusti AG, Barbera JA, Roca J, et al. Hypoxic pulmonary vasoconstriction and gas exchange during exercise in chronic obstructive pulmonary disease. Chest. 1990;97:268–75.

60. Vizza CD, Lynch JP, Ochoa LL, et al. Right and left ventricular dysfunction in patients with severe pulmonary disease. Chest. 1998;113:576–83.
61. Matthay RA, Berger HJ, Davies RA, et al. Right and left ventricular exercise performance in chronic obstructive pulmonary disease: radionuclide assessment. Ann Intern Med. 1980;93:234–9.
62. Arbex FF, Alencar MC, Souza A, et al. Exercise ventilation in COPD: influence of systolic heart failure. COPD. 2016;13(6):693–9.
63. Oliveira MF, Arbex FF, Alencar MC, et al. Heart failure impairs muscle blood flow and endurance exercise tolerance in COPD. COPD. 2016;13(4):407–15.
64. Rocha A, Arbex FF, Alencar MC, et al. Physiological and sensory consequences of exercise oscillatory ventilation in heart failure-COPD. Int J Cardiol. 2016;224:447–53.
65. Watz H. Chronic obstructive pulmonary disease: when pulmonologists do something good for the heart. Am J Respir Crit Care Med. 2016;193:703–4.
66. Stone IS, Barnes NC, James WY, et al. Lung deflation and cardiovascular structure and function in chronic obstructive pulmonary disease. A randomized controlled trial. Am J Respir Crit Care Med. 2016;193:717–26.
67. Morrison DA, Adcock K, Collins CM, et al. Right ventricular dysfunction and the exercise limitation of chronic obstructive pulmonary disease. J Am Coll Cardiol. 1987;9:1219–29.
68. Montes de Oca M, Rassulo J, Celli BR. Respiratory muscle and cardiopulmonary function during exercise in very severe COPD. Am J Respir Crit Care Med. 1996;154(5):1284–9.
69. Furlanetto KC, Mantoani LC, Bisca G, et al. Reduction of physical activity in daily life and its determinants in smokers without airflow obstruction. Respirology. 2014;19:369–75.
70. Regan EA, Lynch DA, Curran-Everett D, et al. Clinical and radiologic disease in smokers with normal spirometry. JAMA Intern Med. 2015;175:1539–49.
71. Woodruff PG, Barr RG, Bleecker E, et al. Clinical significance of symptoms in smokers with preserved pulmonary function. N Engl J Med. 2016;374:1811–21.
72. Elbehairy AF, Guenette JA, Faisal A, et al., Canadian Respiratory Research Network. Mechanisms of exertional dyspnoea in symptomatic smokers without COPD. Eur Respir J. 2016;48(3):694–705.
73. Hogg JC, Chu F, Utokaparch S, et al. The nature of small-airway obstruction in chronic obstructive pulmonary disease. N Engl J Med. 2004;350(26):2645–53.
74. McDonough JE, Yuan R, Suzuki M, et al. Small-airway obstruction and emphysema in chronic obstructive pulmonary disease. N Engl J Med. 2011;365:1567–75.
75. Buist AS. Early detection of airways obstruction by the closing volume technique. Chest. 1973;64:495–9.
76. Barbera JA, Roca J, Ramirez J, et al. Gas exchange during exercise in mild chronic obstructive pulmonary disease. Correlation with lung structure. Am Rev Respir Dis. 1991;144(3 Pt 1):520–5.
77. Rodriguez-Roisin R, Drakulovic M, Rodriguez DA, et al. Ventilation-perfusion imbalance and chronic obstructive pulmonary disease staging severity. J Appl Physiol (1985). 2009;106:1902–8.
78. Kirby M, Owrangi A, Svenningsen S, et al. On the role of abnormal DL(CO) in ex-smokers without airflow limitation: symptoms, exercise capacity and hyperpolarised helium-3 MRI. Thorax. 2013;68:752–9.
79. Harvey BG, Strulovici-Barel Y, Kaner RJ, et al. Risk of COPD with obstruction in active smokers with normal spirometry and reduced diffusion capacity. Eur Respir J. 2015;46(6):1589–97.
80. Elbehairy AF, Faisal A, Guenette JA, et al. Resting physiological correlates of reduced exercise capacity in smokers with mild airway obstruction. COPD. 2017;14(3):267–75.
81. Caviedes IR, Delgado I, Soto R. Ventilatory inefficiency as a limiting factor for exercise in patients with COPD. Respir Care. 2012;57:583–9.
82. Andrianopoulos V, Franssen FM, Peeters JP, et al. Exercise-induced oxygen desaturation in COPD patients without resting hypoxemia. Respir Physiol Neurobiol. 2014;190:40–6.
83. Moreira MA, Medeiros GA, Boeno FP, et al. Oxygen desaturation during the six-minute walk test in COPD patients. J Bras Pneumol. 2014;40:222–8.

84. Patessio A, Casaburi R, Carone M, et al. Comparison of gas exchange, lactate, and lactic acidosis thresholds in patients with chronic obstructive pulmonary disease. Am Rev Respir Dis. 1993;148:622–6.
85. Pleguezuelos E, Esquinas C, Moreno E, et al. Muscular dysfunction in COPD: systemic effect or deconditioning? Lung. 2016;194:249–57.
86. O'Donnell DE, D'Arsigny C, Fitzpatrick M, Webb KA. Exercise hypercapnia in advanced chronic obstructive pulmonary disease: the role of lung hyperinflation. Am J Respir Crit Care Med. 2002;166:663–8.
87. Poon CS, Tin C, Song G. Submissive hypercapnia: why COPD patients are more prone to CO_2 retention than heart failure patients. Respir Physiol Neurobiol. 2015;216:86–93.
88. Sue DY. Excess ventilation during exercise and prognosis in chronic heart failure. Am J Respir Crit Care Med. 2011;183:1302–10.
89. Houben-Wilke S, Jörres RA, Bals R, et al. Peripheral artery disease and its clinical relevance in patients with chronic obstructive pulmonary disease in the COPD and systemic consequences-comorbidities network study. Am J Respir Crit Care Med. 2017;195(2):189–97.
90. O'Donnell DE, Laveneziana P. Dyspnea and activity limitation in COPD: mechanical factors. COPD. 2007;4(3):225–36.
91. O'Donnell DE, Webb KA. Exertional breathlessness in patients with chronic airflow limitation: the role of hyperinflation. Am Rev Respir Dis. 1993;148:1351–7.
92. O'Donnell DE, Hamilton AL, Webb KA. Sensory-mechanical relationships during high-intensity, constant-work-rate exercise in COPD. J Appl Physiol (1985). 2006;101:1025–35.
93. O'Donnell DE, Voduc N, Fitzpatrick M, Webb KA. Effect of salmeterol on the ventilatory response to exercise in chronic obstructive pulmonary disease. Eur Respir J. 2004;24:86–94.
94. O'Donnell DE, Bain DJ, Webb KA. Factors contributing to relief of exertional breathlessness during hyperoxia in chronic airflow limitation. Am J Respir Crit Care Med. 1997;155(2):530–5.
95. O'Donnell DE, D'Arsigny C, Webb KA. Effects of hyperoxia on ventilatory limitation during exercise in advanced chronic obstructive pulmonary disease. Am J Respir Crit Care Med. 2001;163(4):892–8.
96. Somfay A, Porszasz J, Lee SM, Casaburi R. Dose-response effect of oxygen on hyperinflation and exercise endurance in nonhypoxaemic COPD patients. Eur Respir J. 2001;18(1):77–84.
97. Mahler DA, Murray JA, Waterman LA, et al. Endogenous opioids modify dyspnoea during treadmill exercise in patients with COPD. Eur Respir J. 2009;33:771–7.
98. Jensen D, Alsuhail A, Viola R, et al. Inhaled fentanyl citrate improves exercise endurance during high-intensity constant work rate cycle exercise in chronic obstructive pulmonary disease. J Pain Symptom Manag. 2012;43(4):706–19.
99. Johnson M, Bland J, Oxberry S, et al. Opioids for chronic refractory breathlessness: patient predictors of beneficial response. Eur Respir J. 2013;42:758–66.
100. Rocker GM, Simpson AC, Young J, et al. Opioid therapy for refractory dyspnea in patients with advanced chronic obstructive pulmonary disease: patients' experiences and outcomes. CMAJ Open. 2013;1(1):E27–36.
101. Ekström M, Nilsson F, Abernethy AA, Currow DC. Effects of opioids on breathlessness and exercise capacity in chronic obstructive pulmonary disease. A systematic review. Ann Am Thorac Soc. 2015;12(7):1079–92.
102. Carrieri-Kohlman V, Gormley JM, Douglas MK, et al. Exercise training decreases dyspnea and the distress and anxiety associated with it. Monitoring alone may be as effective as coaching. Chest. 1996;110(6):1526–35.
103. Wadell K, Webb KA, Preston ME, et al. Impact of pulmonary rehabilitation on the major dimensions of dyspnea in COPD. COPD. 2013;10(4):425–35.
104. Soumagne T, Laveneziana P, Veil-Picard M, Guillien A, Claudé F, Puyraveau M, Annesi-Maesano I, Roche N, Dalphin JC, Degano B. Asymptomatic subjects with airway obstruction have significant impairment at exercise. Thorax. 2016;71(9):804–11.

Exercise Testing in Cystic Fibrosis

6

Paolo Palange

Abstract

Cystic fibrosis (CF) is the most frequent genetic disease in the Caucasian population. It is characterized by absent or incorrect function of the channel that regulates the chloride exchange at cell surface (CTRF). The lungs are particularly involved as the very thick and tenacious mucus leads to progressive airflow limitation, respiratory infections, bronchiectasis, lung destruction, and ultimately respiratory failure. Dyspnea and exercise intolerance are the hallmarks of the diseases. Patients, however, should be encouraged to exercise regularly since early childhood because it promotes expectoration and contributes to good nutritional status and overall level of fitness. In addition, there are no contraindications to agonist sports. It follows that a growing number of CF patients will be referred to exercise-based evaluations in the forthcoming years. Cardiopulmonary exercise testing (CPET), in particular, is useful to determine the mechanisms of exercise intolerance in individual patients which may have treatment and prognostic implications.

6.1 Introduction

Cystic fibrosis (CF) is the most frequent genetic disease in the Caucasian population due to the absent or incorrect function of the channel that regulates the chloride exchange at cell surface (CTRF). Because of the systemic nature of

P. Palange (✉)
Department of Public Health and Infectious Diseases, Sapienza University of Rome, Rome, Italy

Lorillard-Spencer Foundation, Rome, Italy
e-mail: paolo.palange@uniroma1.it

A. Cogo et al. (eds.), *Exercise and Sports Pulmonology*,
https://doi.org/10.1007/978-3-030-05258-4_6

the disease, many organs are involved; the lungs, however, are mainly affected because of progressive chronic obstructive disease, secondary to the very thick and tenacious mucus that leads to respiratory infections, bronchiectasis, lung destruction, and ultimately respiratory failure. Exercise tolerance is a very well-known prognostic index in CF [1], as it is in many chronic pulmonary and cardiac disorders. Patients with CF should be encouraged to exercise regularly and to have a very active lifestyle from a physical activity point of view, together in keeping a good nutritional state. In CF patients several studies have demonstrated that regular exercise and physical activity together with a good nutritional state are capable to reduce the rate of decline in lung function. Exercise training programs and pulmonary physiotherapy are fundamental therapeutic strategies aimed at improving mucus clearance and at reducing rates of pulmonary exacerbations.

6.1.1 Indications to CPET in CF Patients

It is well known that measurements of lung function, such as FEV_1, obtained at rest provide only a rough estimation of exercise tolerance in patients with CF [2, 3]. Therefore, particularly in the mild and moderate phases of the disease, the evaluation of exercise tolerance should be obtained by using the most appropriate tool and exercise protocol. Cardiopulmonary exercise testing is considered the "gold standard" for the assessment of exercise tolerance [2] in many diseases states. CPET, in addition to clearly define maximal power output (W_{max}) and peak O_2 uptake ($V'O_{2peak}$), provides information of the ventilator response/efficiency ($V'_E/V'CO_2$), of the cardiovascular and metabolic responses to exercise. CF patients with $V'O_{2peak} > 82\%$ predicted normal have more than three times as likely to survive for 8 years than those with low $V'O_{2peak} < 58\%$ predicted [1] (Fig. 6.1). Also, CF patients with a $V'O_{2peak} < 32$ mL/min/kg had an increase in mortality compared to those with $V'O_{2peak} > 45$ mL/min/kg [4]. The following are the indications to CPET in CF (Table 6.1).

6.1.2 CPET Protocols

During CPET measurement of lung gases (O_2 and CO_2), minute ventilation (V'_E), heart rate (HR), arterial blood pressure (BP), and peripheral arterial O_2 saturation (SpO_2) are obtained. Calculated variables include $V'O_2$, carbon dioxide output ($V'CO_2$), breathing reserve (i.e., V'_{Epeak}/maximal voluntary ventilation or MMV), HR reserve (HRR = (220 − age) − HR_{peak}), and O_2 pulse (O_2 pulse = $V'O_2$/HR). Important additional information is provided with measurements of symptoms (dyspnea and leg fatigue), inspiratory capacity (IC), arterial blood gases (PaO_2 and $PaCO_2$), and blood lactate (La-) obtained at rest, during and at peak of exercise. In adult subjects the maximal incremental ramp test on a cycle ergometer is the protocol recommended by the ERS [2]; the test is expected to last 10–12 min during

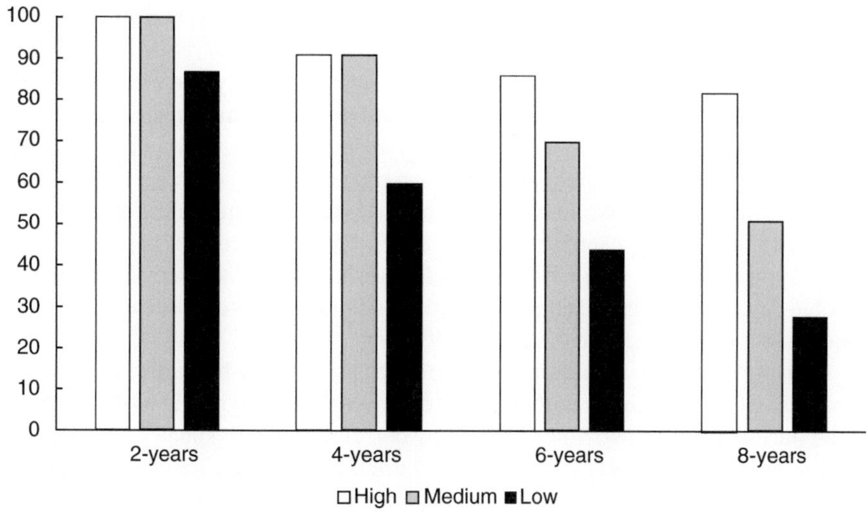

Fig. 6.1 Rate of survival (%) among CF patients according to aerobic fitness (i.e., maximal aerobic capacity). Level of fitness according to maximal aerobic capacity ($V'O_{2peak}$) measured during cardiopulmonary exercise testing (CPET). High, $V'O_{2peak} \geq 82\%$ predicted; medium, $V'O_{2peak} = 81 - 59\%$ predicted; low, $V'O_{2peak} \leq 58\%$ predicted (According to data from Ref. [1])

Table 6.1 Indications to CPET in CF

1. Functional and prognostic evaluation.
2. Evaluation of the causes of exercise intolerance.
3. Detection of exercise-induced bronchoconstriction.
4. Detection of exercise-induced arterial oxygen desaturation.
5. Preoperative evaluation, e.g., lung transplantation.
6. Exercise prescription before exercise-based rehabilitation.
7. Evaluation of therapeutic interventions, e.g., drugs, rehabilitation.

which subject is pushed to his/her limit of tolerance. In CF children, the maximal incremental Godfrey cycle ergometer protocol [4] has been utilized extensively. It requires the subject to maintain a rate of 60 revolutions per minute while the load is increased every minute until volitional fatigue. Treadmill exercise protocols have been widely used in clinical assessment of myocardial ischemia in adult subjects, being the Bruce protocol [5] the most utilized; this protocol has been utilized to detect changes with interventions in children with CF [6]. The Bruce protocol requires the subject to walk initially and run subsequently on a treadmill, where both the speed and the gradient of incline are increased every 3 min. It should be kept in mind that power output during a treadmill test is estimated, while with cycle ergometer exercise, a precise measurement of power output is obtained if the ergometer is calibrated routinely; if measurement of work efficiency (i.e., $V'O_2/W$) is required, the cycle ergometer or another calibrated ergometer (e.g., arm ergometer) should be used.

6.1.3 Factors Limiting Exercise Tolerance in CF Detected at CPET

In the advanced phase of the disease, CF patients are usually exercise limited by ventilatory constraints (e.g., low FEV_1). However, "peripheral factors" related to deconditioning and/or to poor nutritional state may play a very important role. Dyspnea and leg fatigue are usually the symptoms limiting exercise tolerance in CF. Interestingly, as shown in a recent published study, the frequency of these two symptoms in stopping exercise does not differ, CF versus control [7]. Poor skeletal muscle condition, with reduced muscle mass and force, is associated with low aerobic capacity and early lactate production; this may lead to increase ventilatory requirement during exercise (i.e., higher V'_E/W).

As in other chronic respiratory diseases such as chronic obstructive pulmonary diseases (COPD), patients with CF may experience lung dynamic hyperinflation (DH) during exercise [8]. Since DH highly correlates with dyspnea sensation and is susceptible to treatment with bronchodilators, this negative mechanism should be ruled out during exercise testing through measurements of IC and end expiratory lung volume (EELV). Arterial oxygen desaturation may be observed during exercise. In other diseases such as COPD and lung fibrosis, arterial oxygen desaturation usually occurs when diffusion lung CO (DLco) is < 50% of predicted; also, arterial oxygen desaturation is more evident during walking compared to cycling exercise [9]. In some CF patients, due to the deconditioning of peripheral muscles and poor muscle energy, anaerobic metabolism with lactate production occurs at the beginning of the exercise. The anaerobic contribution to exercise metabolism (in addition to the aerobic one) can be detected noninvasively by the use of the lactate threshold (LT). The recommended graphical method to detect LT is the "V-slope" originally described by Beaver et al. [10]; by plotting exercise $V'CO_2$ and $V'O_2$, data is possible to detect a change in the slope of the relationship that indicates LT. In normal subjects, LT occurs at 40–50% of the predicted VO_{2max}. Ultimately CPET helps in determining not only the degree of exercise limitation (i.e., $V'O_{2peak}$) that predicts prognosis in CF [1] but also gives a better insight on the causes of the reduced exercise tolerance. The possible causes of exercise intolerance (i.e., low $V'O_{2peak}$) detectable at CPET in CF are illustrated in Table 6.2.

Table 6.2 Possible causes of exercise intolerance in CF

Cause	CPET finding
1. Reduced ventilatory capacity.	Low breathing reserve; high $V'_E/V'O_2$
2. Deconditioning, poor nutritional state.	Early LT, high $HR/V'O_2$
3. Exercise-induced bronchoconstriction.	>10% drop in baseline FEV_1
4. Exercise-induced arterial oxygen desaturation.	>4% drop in baseline SpO_2
5. Pulmonary hypertension (rare).	$V'O_2/W$ slope < 8
	High $V'_E/V'CO_2$ (slope and nadir)
	Low O_2 pulse; high $HR/V'O_2$

6.1.4 Field Tests

When exercise lung gas exchange equipment are not available, field tests can be utilized. These tests are however less sensitive than CPET in assessing the degree of exercise intolerance and the responses to treatment, compared to the constant work exercise protocols at CPET [11]. The distance covered at the 6-min walking test (6 MWT) or the number of steps climbed during a 3-min step test [12] has been utilized in CF studies to evaluate the response to therapeutic intervention. It is recommended to measure HR and SpO_2 at baseline and at peak exercise. While the 6 MWT is a self-paced walking test that roughly correlates with $V'O_{2peak}$ measured at CPET, the "shuttle test" is an incremental externally paced audio signal. The highest speed measured at shuttle test better correlates, compared to 6 MWT, with $V'O_{2peak}$ measured at CPET [13].

6.1.5 Pulmonary Exercise-Based Rehabilitation in CF

Patients with CF should be encouraged to exercise regularly since the early years of life. It has been demonstrated that even moderate exercise promotes sputum expectoration by increasing airway lining fluid [14, 15]. Exercise-based rehabilitation programs clearly demonstrated beneficial effects in CF patients [16–18]. In addition, exercise-based rehabilitation programs have demonstrated to reduce the rate of pulmonary exacerbations [19]. A Cochrane review is available that demonstrates definitive benefit of exercise programs in subjects with CF [20].

6.1.6 Daily Physical Activity in CF

In normal individuals, the levels of daily physical activity (PA) and maximal exercise tolerance are somewhat related. Recent studies from our laboratory have demonstrated that in CF levels of PA are better evaluated by the use of accelerometers compared to questionnaires [21]. In addition, we were able to demonstrate that daily PA positively correlated with maximal aerobic fitness (i.e., $V'O_{2peak}$) [22]. Perhaps more importantly, adult CF patients with more pulmonary exacerbations in the preceding year have more advanced disease and are less active than their peers; in this study PA was independently associated with gender and airflow obstruction, being the females less active than males [23].

Key Points

- Cystic fibrosis (CF) is a genetic disease due to the absent or incorrect function of the channel that regulates the chloride exchange at cell surface (CTRF). Lungs are mainly affected because of progressive chronic obstructive disease, secondary to the very thick and tenacious mucus that leads to respiratory infections, bronchiectasis, lung destruction, and ultimately respiratory failure.

- Exercise intolerance is a hallmark of the disease. Measurements of lung function, such as FEV_1, obtained at rest provide a rough estimation of exercise tolerance in CF patients. Particularly in the mild and moderate phases of the disease, the evaluation of exercise tolerance should be obtained by using the most appropriate tool and exercise protocol, such as cardiopulmonary exercise testing (CPET).
- In the advanced phase of the disease, CF patients are usually exercise limited by ventilatory constraints. Also peripheral factors related to deconditioning and poor nutritional state may play a very important role. Dyspnea and leg fatigue are usually the symptoms limiting exercise tolerance in CF.
- In CF regular exercise since early childhood promotes expectoration and contributes to good nutritional status and overall level of fitness, which, in turn, are capable to reduce the rate of decline in lung function.
- Cardiopulmonary exercise testing (CPET), in particular, is useful to determine the mechanisms of exercise intolerance in individual patients which may have treatment and prognostic implications.

References

1. Nixon PA, Orenstein DM, Kelsey SF, Doershuk CF. The prognostic value of exercise testing in patients with cystic fibrosis. N Engl J Med. 1992;327(25):1785–8.
2. Clinical exercise testing with reference to lung diseases: indications, standardization and interpretation strategies. ERS Task Force on Standardization of Clinical Exercise Testing. European Respiratory Society. Eur Respir J. 1997;10(11):2662–89.
3. ERS Task Force, Palange P, Ward SA, Carlsen KH, Casaburi R, Gallagher CG, Gosselink R, O'Donnell DE, Puente-Maestu L, Schols AM, Singh S, Whipp BJ. Recommendations on the use of exercise testing in clinical practice. Eur Respir J. 2007;29(1):185–209.
4. Pianosi P, LeBlanc J, Almudevar A. Relationship between FEV1 and peak oxygen uptake in children with cystic fibrosis. Pediatr Pulmonol. 2005;40(4):324–9.
5. Bruce RA, Kusumi F, Hosmer D. Maximal oxygen intake and nomographic assessment of functional aerobic impairment in cardiovascular disease. Am Heart J. 1973;85(4):546–62.
6. Robinson PD, Cooper P, Van Asperen P, Fitzgerald D, Selvadurai H. Using index of ventilation to assess response to treatment for acute pulmonary exacerbation in children with cystic fibrosis. Pediatr Pulmonol. 2009;44(8):733–42.
7. Quon BS, Wilkie SS, Ramsook AH, Schaeffer MR, Puyat JH, Wilcox PG, Guenette JA. Qualitative dimensions of exertional dyspnea in adults with cystic fibrosis. J Appl Physiol (1985). 2016;121(2):449–56.
8. Nicholson TT, Barry PJ, Waterhouse DF, Nolan GM, McKone EF, Gallagher CG. Relationship between pulmonary hyperinflation and dyspnoea severity during acute exacerbations of cystic fibrosis. Respirology. 2017;22(1):141–8.
9. Palange P, Forte S, Onorati P, Manfredi F, Serra P, Carlone S. Ventilatory and metabolic adaptations to walking and cycling in patients with COPD. J Appl Physiol (1985). 2000;88(5):1715–20.

10. Beaver WL, Wasserman K, Whipp BJ. A new method for detecting the anaerobic threshold by gas exchange. J Appl Physiol. 1986;60:2020–7.
11. Puente-Maestu L, Palange P, Casaburi R, Laveneziana P, Maltais F, Neder JA, O'Donnell DE, Onorati P, Porszasz J, Rabinovich R, Rossiter HB, Singh S, Troosters T, Ward S. Use of exercise testing in the evaluation of interventional efficacy: an official ERS statement. Eur Respir J. 2016;47(2):429–60.
12. Kazmerski T, Orenstein DM. The ease of breathing test tracks clinical changes in cystic fibrosis. J Cyst Fibros. 2012;11(5):383–6. https://doi.org/10.1016/j.jcf.2012.03.005.
13. Onorati P, Antonucci R, Valli G, Berton E, De Marco F, Serra P, Palange P. Non-invasive evaluation of gas exchange during a shuttle walking test vs. a 6-min walking test to assess exercise tolerance in COPD patients. Eur J Appl Physiol. 2003;89(3–4):331–6.
14. Schmitt L, Wiebel M, Frese F, Dehnert C, Zugck C, Bärtsch P, Mairbäurl H. Exercise reduces airway sodium ion reabsorption in cystic fibrosis but not in exercise asthma. Eur Respir J. 2011;37(2):342–8.
15. Salh W, Bilton D, Dodd M, Webb AK. Effect of exercise and physiotherapy in aiding sputum expectoration in adults with cystic fibrosis. Thorax. 1989;44(12):1006–8.
16. Orenstein DM, Hovell MF, Mulvihill M, Keating KK, Hofstetter CR, Kelsey S, Morris K, Nixon PA. Strength vs aerobic training in children with cystic fibrosis: a randomized controlled trial. Chest. 2004;126(4):1204–14.
17. Klijn PH, Oudshoorn A, van der Ent CK, van der Net J, Kimpen JL, Helders PJ. Effects of anaerobic training in children with cystic fibrosis: a randomized controlled study. Chest. 2004;125(4):1299–305.
18. Moorcroft AJ, Dodd ME, Morris J, Webb AK. Individualised unsupervised exercise training in adults with cystic fibrosis: a 1 year randomised controlled trial. Thorax. 2004;59(12):1074–80.
19. Schneiderman-Walker J, Pollock SL, Corey M, Wilkes DD, Canny GJ, Pedder L, Reisman JJ. A randomized controlled trial of a 3-year home exercise program in cystic fibrosis. J Pediatr. 2000;136(3):304–10.
20. Bradley J, Moran F. Physical training for cystic fibrosis. Cochrane Database Syst Rev. 2008;1:CD002768.
21. Savi D, Quattrucci S, Internullo M, De Biase RV, Calverley PM, Palange P. Measuring habitual physical activity in adults with cystic fibrosis. Respir Med. 2013;107(12):1888–94.
22. Savi D, Di Paolo M, Simmonds N, Onorati P, Internullo M, Quattrucci S, Winston B, Laveneziana P, Palange P. Relationship between daily physical activity and aerobic fitness in adults with cystic fibrosis. BMC Pulm Med. 2015;15:59.
23. Savi D, Simmonds N, Di Paolo M, Quattrucci S, Palange P, Banya W, Hopkinson NS, Bilton D. Relationship between pulmonary exacerbations and daily physical activity in adults with cystic fibrosis. BMC Pulm Med. 2015;15:151.

Exercise in Interstitial Lung Diseases

7

Baruch Vainshelboim

Abstract

Interstitial lung diseases (ILDs) represent a group of diverse chronic lung conditions characterized by scaring, inflammation, and restrictive pathophysiology. Idiopathic pulmonary fibrosis (IPF) is the most common form of ILDs and is associated with severe signs and symptoms, exercise intolerance, impaired quality of life, and poor prognosis. Exercise training has been shown to be a safe and effective treatment in a variety of ILDs including IPF. Exercise training improves exercise tolerance, functional capacity, dyspnea, and quality of life both in patients with ILDs and IPF. Higher exercise capacity and physical activity levels have been demonstrated to be associated with better survival and enhanced quality of life. The optimal training modalities and the underlining mechanisms with respect to outcome improvements are yet to be well characterized and require further investigation. The existing evidence supports the clinical benefits of exercise in patients with ILDs and thus provides good justification to recommend exercise training as part of standard care for ILDs.

7.1 Introduction

Interstitial lung diseases (ILDs), also known as diffuse parenchymal lung diseases, encompass a wide and diverse group of more than 200 disorders affecting not only the interstitium but also peripheral airways, alveoli, and small blood

B. Vainshelboim (✉)
Master of Cancer Care Program, School of Health Sciences, Saint Francis University, Loretto, PA, USA

Pulmonary Institute, Rabin Medical Center, Beilinson Hospital, Petach Tikva, Israel

Faculty of Sport, Research Center in Physical Activity, Health and Leisure (CIAFEL), University of Porto, Porto, Portugal

A. Cogo et al. (eds.), *Exercise and Sports Pulmonology*,
https://doi.org/10.1007/978-3-030-05258-4_7

vessels within the lungs. ILDs are characterized by lung inflammation, scarring, and presence of restrictive pathophysiology. ILDs include diseases of underlying systemic processes (i.e., sarcoidosis), connective tissue diseases (i.e., rheumatoid arthritis), occupational exposures (i.e., asbestosis, silicosis), collagen vascular diseases, as well as environmental and drug-related diseases. ILDs also include conditions of unknown etiologies such idiopathic interstitial pneumonia or idiopathic pulmonary fibrosis (IPF). IPF is the most common form of ILDs and have no established etiology [1, 2].

IPF is a chronic, progressive ILD associated with high morbidity and a median survival of 2–5 years from the time of diagnosis [3–5]. IPF is characterized by progressive pulmonary restriction, ventilatory inefficiency, dyspnea, impaired gas exchange, and hypoxemia, which all lead to diminished exercise capacity [3, 5, 6]. Patients with IPF experience severe breathlessness and tend to be less physically active in order to avoid such symptoms [7, 8]. Manifestations of IPF have a negative impact on functional capacity and quality of life (QOL) and are associated with poorer prognosis [3, 5–8]. Long-term effective treatment apart from lung transplantation is still limited for most IPF patients despite some encouraging recent findings with pharmacotherapies [3, 9–11].

A recent systematic review demonstrated that short-term exercise training in pulmonary rehabilitation settings is a safe and effective treatment for improving exercise capacity, dyspnea, and quality of life in patients with ILDs and IPF [12]. The current chapter comprehensively reviews the existing evidence on physical activity and rehabilitation programs in patients with ILDs and IPF, highlighting important insights concerning exercise in the management of these diseases.

7.2 Pathophysiology

At rest, patients with ILDs and IPF frequently show a restrictive pulmonary physiology characterized by reduced forced vital capacity (FVC) and total lung capacity (TLC), combined with impaired gas exchange [5, 6]. The resting arterial blood gases are usually near normal or may reveal mild hypoxemia; however, breathing pattern is often rapid and shallow. In general, as the disease progresses, lung compliance decreases and lung volumes fall [6]. Dyspnea is a predominant symptom of ILDs and IPF and results clinically significant due to its strong association with exercise intolerance, poor quality of life, and mortality [3, 8, 13–15]. Other symptoms such as leg pain, chest discomfort, and fatigue are also common reasons for exercise test termination [16, 17]. Patients with IPF can be also bothered by a dry cough which interferes with daily activities. The onset of symptoms is slow, despite progressive [5]. Exercise intolerance is a cardinal feature in ILDs and is associated with severe exertional dyspnea and fatigue, as well as poor quality of life [16–18]. Patients with IPF often exhibit reduced peak aerobic capacity (VO_2peak), peak work rate, and sub-maximal exercise endurance (anaerobic threshold) compared to age- and sex-matched healthy controls [16].

7.3 Exercise Tolerance Assessment

7.3.1 Cardiopulmonary Exercise Testing

Cardiopulmonary exercise testing (CPET) is the gold-standard procedure to objectively assess cardiorespiratory fitness (VO_2 peak/VO_2 max), ventilatory, electrocardiographic, and metabolic responses to exercise [19–24]. Treadmill and cycle ergometers are the most commonly adopted CPET protocols among patients with chronic heart and lung diseases. CPET provides a comprehensive evaluation of exercise tolerance and functional capacity, detecting limiting factors during exercise; it also aids in establishing diagnosis and in reliably assessing responses to interventions [19–24]. In addition, variables obtained from CPET provide powerful prognostic information that has been shown to be more sensitive than traditional measures among patients with cardiopulmonary diseases [21–23]. Table 7.1 shows prognostic markers in patients with IPF.

Table 7.1 Selected prognostic markers in ILDs and IPF

Variables	Studies/thresholds	Hazard ratio (95% confidence intervals) for mortality
Low 6-min walk distance	du Bois et al. 2014 [25]<250 m Kawut et al. 2005 [26]<350 m Lederer et al. 2006 [27]<207 m Caminati et al. 2009 [28]<212 m	2.1 (1.2–3.9) 4.6 (1.5–14.2) 4.7 (2.5–8.9) Not reported
Desaturation during 6-min walk test	Lama et al. 2003 [29] $SpO_2 < 88\%$ Vainshelboim et al. 2016 [30] Δ $SpO_2 \geq 10\%$	4.5 23.3 (1.5–365)
Low exercise capacity	Fell et al. 2009 [31] VO_2 peak <8.3 mL/kg/min Triantafillidou et al. 2013 [32] VO_2 peak >14.2 mL/kg/min per one unit increase Vainshelboim et al. 2016 [33] work rate < 62 watts Layton et al. 2017 [34] peak work rate %predicted<35%	3.2 (1.1–9.6) 0.75 (0.6–0.95) 9.2 (1.9–42.6) 4.7 (2.6–8.4)
Ventilatory inefficiency	Triantafillidou et al. 2013 [32] VE/VCO_2 at anaerobic threshold per one unit increase Vainshelboim et al. 2016 [33] VE/VCO_2 at anaerobic threshold>34	1.15 (1.04–1.26) 4.6 (1.2–17.3)
Physical activity	Vainshelboim et al. 2016 [30] ≤417 METS-min/week	9.7 (1.3–72)
	Vainshelboim et al. 2018 [35]	
	Walking 100 to <150 vs. <100 min/week	0.38 (0.16–0.88)
	Walking ≥150 vs. <100 min/week	0.14 (0.03–0.53)
Prolonged sitting time	Vainshelboim et al. 2018 [35]	
	5 to <10 vs. <5 h/day	4.6 (1.3–16.3)
	≥10 vs. <5 h/day	21.2 (4.1–32.6)

Reviews and research studies utilizing CPET in ILDs usually show multifactorial limitations during exercise. These include abnormal pulmonary gas exchange, inefficient breathing mechanics, exercise-induced hypoxemia, circulatory impairments, and respiratory and skeletal muscle dysfunctions [6, 16, 17, 36]. A hallmark clinical sign in patients with ILDs is hypoxemia, which manifests as a decline in arterial O_2 pressure and arterial O_2 saturation in response to exercise. This phenomenon mainly relates to abnormalities in pulmonary gas exchange, due to alveolar ventilation-perfusion (VA/Q) mismatching, oxygen diffusion limitation, and low mixed venous oxygen content [16, 17]. Ventilatory pattern is also seen abnormal in patients with ILDs; however breathing reserve in most cases is kept within normal values [16, 17]. In particular, part of the raised ventilatory drive during exercise is related to the increased dead space ventilation, which may also relate to underlying pulmonary vascular diseases, especially chronic pulmonary emboli or associated emphysema [6, 16, 17].

7.3.2 6-Min Walk Test

The 6-min walk test (6MWT) is a well-established test for assessing functional capacity in elderly subjects and in patients with cardiopulmonary diseases [37]. The primary outcome of the 6MWT is the 6-min walk distance (6MWD), a metric that provides a valid and reliable estimate of exercise capacity for people with chronic lung diseases. The 6MWD is associated with peak work rate capacity, physical activity, and VO_2 peak, supporting its role as a functional endpoint of exercise performance [38, 39]. The 6MWD has a good sensitivity to detect significant changes in pre-post exercise-based rehabilitation interventions. The minimal clinical important difference for the 6MWD ranges from 25 m to 33 m in adults with chronic respiratory diseases, and most of the trials performed in ILDs and IPF patients show clinical improvement by meeting or exceeding this threshold [12, 18, 38–40]. Desaturation during 6MWT in patients with ILDs and especially IPF has been shown to be an important clinical sign. The nadir SpO_2 during 6MWT aids in determining disease severity and providing prognostic values, such as the need for lung transplantation [3, 29]. The distance completed during the 6MWT has been also consistently proven to be associated with mortality among patients with IPF and has been used as an important marker of Lung Allocation Score for lung transplantation (Table 7.1) [25–28, 41].

7.4 Exercise Training in ILDs

In the past decade, growing body of evidence revealed the safety and efficacy of exercise training interventions for patients with ILDs and IPF [15, 42–55]. This was confirmed by a recent Cochrane systematic review and meta-analysis from Dowman et al. [12]. The analysis included five randomized controlled trials (86 subjects who took part in pulmonary rehabilitation programs and 82 controls). Significant improvements in 6MWD [44.3 m 95% CI (26.04–62.64)] and in VO_2 peak [1.24 mL/kg/min 95% CI (0.46–2.03)] were shown for all ILD patients. Similar findings were reported in the subgroup of IPF patients with significant improvements of 36 m, 95% CI [16–55], and 1.5 mL/kg/min, 95% CI (0.5–2.4), respectively. The results from this meta-analysis also showed a reduced dyspnea and improved quality of life both in ILD and IPF patients with comparable efficacy [12, 56]. These findings are consistent with the previous Cochrane review by Holland et al. demonstrating clinical benefits of exercise training in ILD and IPF [56]. Of note, both meta-analyses showed an average improvement of 6MWD exceeding the minimal clinical important difference of 25 m generally set for chronic respiratory patients [12, 18, 38, 39, 57].

In general, exercise programs were undertaken for 4–12 weeks utilizing 2–3 weekly exercise sessions of 30–60 min duration. All training programs included aerobic endurance exercises, such as walking and cycling, whereas some programs combined also resistance and flexibility training, respiratory muscle training, and breathing exercises [16, 52].

7.5 Skeletal Muscle Strength and Endurance

Muscular strength and endurance are important health-related fitness components due to their associations with functional capacity and daily life activities [58, 59]. However, only few studies of exercise training investigated the strength and endurance of peripheral muscles in patients with ILDs [53, 54]. Available findings showed beneficial effect of exercise training on muscular strength and endurance. In particular, Arizono et al. showed significant improvements in handgrip and quadriceps strength following 10 weeks of a pulmonary rehabilitation program in patients with IPF [53]. Vainshelboim et al. [54] demonstrated a significant improvement in a 30-s chair-stand test (suggestive of the functional strength-endurance capacity of lower limb muscles) after 12-week exercise program in patients with IPF [54]. However, the data are still limited, and future studies are needed to fully address the muscular fitness component in ILDs and IPF patients.

7.6 Physical Activity

Physical inactivity has been identified by the World Health Organization as the fourth leading risk factor for global mortality [60]. Approximately 31% of adults are estimated to be physically inactive, contributing to 6% of all deaths [60, 61]. Physical activity has not been studied extensively in patients with ILDs. The manifestation of ILDs and the presence of signs and symptoms, especially during physical exertion, may mechanistically explain the high prevalence of inactivity in order to avoid breathless, fatigue, and other related respiratory symptoms [5, 7]. In general, inactivity among chronic respiratory disease patients is associated with poorer outcomes including higher mortality risk [18].

Few observational studies among patients with IPF have shown that low physical activity levels are associated with higher mortality rates [30, 62–64]. The study of Wallaert et al. [63], by using accelerometers for step counting, found a 65% lower daily physical activity in patients with IPF compared to healthy sedentary controls [63]. This study also showed that among these patients, physical activity <3287 steps/day was associated with poorer survival rates [63]. In align, Nakayama et al. [62] showed that low physical activity was associated with disease severity, as measured by blood biomarkers, extent of honeycombing, 6MWD, and dyspnea levels among stable IPF patients [62]. Vainshelboim et al. [30] using an international physical activity questionnaire reported that physical activity levels ≤417 METS-min/week (corresponding to 100–105 min of moderate-intensity physical activity per week) were associated with almost 10 times higher risk for mortality during a 40-month follow-up in patients with IPF [30]. More recently, Bahmer et al. [64] relying on step counters accelerometers showed that physically active patients with IPF exhibited significant lower risk of mortality (hazard ratio = 0.46) compared to inactive individuals during a median of 34-month follow-up [64].

Nonetheless, the effects of participating in supervised exercise training or pulmonary rehabilitation programs on physical activity were not extensively studied among patients with ILDs and IPF. In fact, only three prospective studies examined the effect of short-term exercise-based pulmonary rehabilitation programs on physical activity levels in IPF, with some conflicting results in the follow-up reassessment [50, 65, 66]. These studies consistently reported a short-term increase in physical activity after completion of the exercise-based interventions [50, 65, 66]. However, during follow-up while Gaunaurd [66] and Vainshelboim [65] showed reduction in physical activity levels, assessed by the international physical activity questionnaire, Ryerson [50] demonstrated conservation of physical activity levels, based on the rapid assessment of physical activity questionnaire scores. Interestingly, while the average physical activity levels in Vainshelboim's study [65] declined during the 11-month follow-up, at individual level 57% of the patients who participated in the exercise training were still above the baseline with preserved minimal clinically important difference of 200 METs-min/week (50 min/week at moderate intensity). This may provide a preliminary support for long-term positive effects of supervised exercise programs on home-based physical activity [65].

Taken together, preliminary data suggest benefits and better survival in active compared to inactive patients with IPF. Limited data also suggest the potential for pulmonary rehabilitation and supervised exercise programs to increase activity levels among patients with IPF. However, the existing evidence is confined to small sample sizes and lacks validation with objective devices (step counters/accelerometers), thus requiring further research. In addition, large prospective cohorts are warranted to characterize the role of physical activity in ILDs prognosis, and the long-term effects of exercise interventions on physical activity need to be explored more in depth.

7.7 Sedentary Behaviors

Population-based studies have found that more than half of an average person's waking hours involve sedentary activity, mainly prolonged sitting, such as watching television and using computer [67]. Prolonged sitting time was also recently acknowledged as independent risk factor from physical inactivity for incidence of many chronic diseases, hospitalizations, and mortality [68, 69]. While sedentary behaviors got little scientific attention in patient with ILDs and IPF, few available data demonstrated that sedentary lifestyle were associated with poorer outcomes [35, 70, 71]. Atkins et al. [70] found a trend toward significant association between prolonged sitting and mortality at 1 and 2 years of follow-up in IPF patients [70]. Vainshelboim et al. [35] showed that compared to patients with IPF who reported sitting time < 5 h/day, patients who sat ≥5 h/day experienced an increased risk of hospitalizations and mortality [35]. These preliminary findings support the potential clinical significance of sedentary behaviors in patients with IPF that need to be further addressed in future large prospective studies.

7.8 Possible Physiological Mechanisms of Exercise Training Effect in ILDs

ILDs are complex chronic lung diseases which result in intra- and extrapulmonary impairments that often worsen over time [1–3, 5, 16, 17]. Exercise training in healthy population has been shown to positively impact on cardiovascular, respiratory, and musculoskeletal systems [72]. Chronic appropriate stimuli with exercise may result in beneficial training effects and physiological adaptation for patients with ILDs, despite the existence of pathophysiological abnormalities and impairments [16, 17, 59]. Figure 7.1 illustrates possible mechanisms for the beneficial effects of exercise training in patients with ILDs.

A randomized controlled study by Vainshelboim et al. [54] revealed a significant improvement in ventilatory functions, including peak minute ventilation and peak tidal volume, after 12 weeks of an exercise training intervention in patients with IPF [54]. The improvement in peak tidal volume was significantly correlated ($r = 0.78$, $p = 0.001$) with an improvement in VO_2peak values, a gold-standard marker of cardiorespiratory fitness associated with better survival and overall health [19, 22, 23, 54, 73]. This underlying mechanism is poorly understood but may be related to a repetitive stimulus of high ventilatory demands, chest expansion during breathing exercises and stretching of the thoracic muscles during exercise sessions. These could result in a more efficient breathing pattern, improved strength of respiratory muscles, enhanced pleural elasticity, and pulmonary compliance [6, 16, 54, 74]. The abovementioned is consistent with a review paper suggesting beneficial effects of thoracic expansion and stretching on pulmonary restriction for IPF [74]. In addition, patients included in the study of Vainshelboim et al. [54] also showed improvements

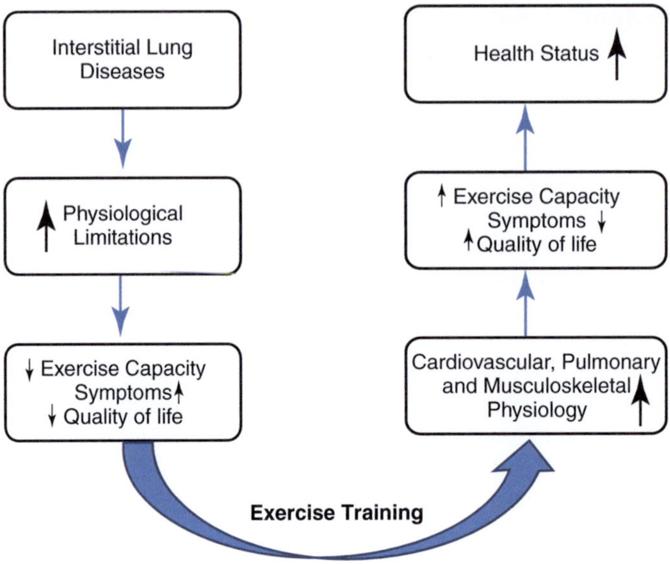

Fig. 7.1 Potential physiological mechanisms of exercise training in patients with interstitial lung diseases

in dyspnea, a phenomenon that further supports the relationship between cardiorespiratory fitness, ventilatory capacities, and exertional dyspnea [54]. These findings are consistent with data provided by Manali et al. [75] who found a significant correlation between dyspnea and VO_2peak in patients with IPF [75]. It also may be that the improvement in exercise capacity and ventilatory function results in reduced dyspnea at sub-maximal exercises, such as activities of daily living, as demonstrated by declined dyspnea scale rating after the program. This enhancement could determine an increase in alveolar oxygen tension and improvements in alveoli ventilation/perfusion (VA/Q) mismatch, resulting in an improved VO_2peak [5, 6, 16].

Furthermore, the study conducted by Keyser et al. [76] found significant peripheral adaptation after 10-week aerobic treadmill exercise program among ILD patients [76]. The authors used a near-infrared spectroscopy of peripheral oxygen extraction, suggesting it as primary physiological mechanism of increase in aerobic capacity in ILD [76]. In addition, Vainshelboim et al. [77] found a significant improvement in cluster of noninvasive exercise cardiovascular indexes, representing cardiac power and heart contractility among patients with IPF. This improvement was significantly correlated with improvements in functional capacity (6MWD) [77]. Given a considerably high (66%) prevalence of coronary artery disease in patients with IPF, the improvement in exercise cardiovascular function is potentially clinically important for cardiac disease prognosis and risk reduction in patients with IPF [77, 78].

Taken together, physiological studies in patients with ILDs and IPF suggest a significant improvement in ventilatory, cardiovascular, and skeletal muscle functions. However, the exact underlying mechanisms of training adaption in patients with ILDs and IPF are yet to be understood and require future research utilizing low-dose computed tomography, stress echocardiography, and near-infrared spectroscopy to clearly address the intra- and extrapulmonary anatomical and physiological adaptations to exercise interventions.

7.9 Conclusions

Interstitial lung diseases, especially IPF, are associated with significant morbidity and mortality, exercise intolerance, dyspnea, and poor quality of life. Exercise training is feasible, safe, and clinically effective for improving exercise capacity, dyspnea, and quality of life in patients with ILDs and IPF. Higher exercise capacity and physical activity levels in patients with IPF were associated with better survival. Long-term effects of exercise training and fitness, as well as the physiological mechanisms, are yet to be determined.

7.10 Tips and Pitfalls

Despite the emerging scientific evidence with respect to safety and efficacy of exercise training in patients with ILDs and IPF, specific set of guidelines are yet to be published [74, 79]. Although exercise training studies in patients with ILDs show a good safety profile, these patients tend to have additional comorbidities, such as coronary arterial disease, systemic and pulmonary hypertension, as well as significant symptoms during exercise [3, 5, 6, 16, 78]. Thus, baseline comprehensive evaluation would be beneficial before starting an exercise training program. This may include a respiratory physician evaluation, pulmonary function test, 6MWT, "30-second chair sit to stand test," and "8-foot up and go test" [18, 23]. In addition, ILD patients present multifactorial limitations during exercise; integrative CPET would be therefore valuable to assess the electrocardiographic, hemodynamic, respiratory, and gas exchange responses to exercise [16, 17]. CPET can also serve to optimize personal training program prescriptions and necessary adjustments in oxygen supplementation during exercise sessions [18].

Oxygen supplementation seems to be necessary for hypoxemic patients during exercise as has been recommended by the American Thoracic Society/European Respiratory Society guidelines [3]. $SpO_2 > 90\%$ was previously recommended as appropriate oxygenation threshold for ILD patients [80], based on COPD data [81]. However, ILD patients tend to desaturate to much lower levels compared to COPD patients and practically may require different thresholds for oxygen supplementation delivery [82, 83]. In this regard, SpO_2 thresholds $<88\%$ [54] and $<85\%$ [49] were previously reported for oxygen supplementation during exercise sessions and seem to be reasonable and safe [49, 54]. A specific set of guidelines on oxygen supplementation during exercise in ILDs are yet to be published, and suggestions provided in this chapter are based on observational reports [15, 42–50, 54] and the pulmonary rehabilitation documents of leading respiratory organizations [18, 84, 85].

7.11 Practical Recommendations for Supervised Exercise Training Program

- Identify an appropriate space for conducting the exercise training program.
- Rely on staff specialized in exercise rehabilitation and familiar with signs and symptoms of ILDs.
- Address safety requirements in terms of emergency plans, oxygen delivery, and necessary monitoring during exercise sessions.
- Assess patients at baseline and after the intervention with at least medical evaluation, 6MWT, "30-second chair sit to stand test," "8-foot up and go test," dyspnea scale, and QOL questionnaire.
- Deliver at least two supervised exercise sessions per week over 6–12 weeks (longer is better).
- Include in exercise sessions 20–45 min of aerobic exercise (can start using an interval training of 5 min*4–6 bouts with 1 min rest between the bouts), 15–25 min of resistance/strength activities, and 10–15 min of flexibility and breathing exercises.
- Close monitoring of SpO_2, heart rate, blood pressure, Borg dyspnea scale, and clinical symptoms during exercise sessions.
- Incorporate educational sessions and psychological support with the exercise training as recommended for comprehensive pulmonary rehabilitation program.
- Encourage patients for home-based physical activity on other days of the week and consult to reduce sedentary behaviors.
- Develop a maintenance plan in order to sustain the improvement in outcomes.

Key Points

- Interstitial lung diseases (ILDs), especially idiopathic pulmonary fibrosis (IPF), are associated with significant morbidity, mortality, exercise intolerance, dyspnea, and poor quality of life.
- Interventional studies demonstrate that exercise training is feasible, safe, and clinically effective for improving exercise capacity, dyspnea, and quality of life in patients with ILDs.
- Preliminary observational data show that higher exercise capacity and physical activity levels are associated with better survival in IPF.

References

1. Kebbe J, Abdo T. Interstitial lung disease: the diagnostic role of bronchoscopy. J Thorac Dis. 2017;9(Suppl 10):S996–S1010.
2. Holland AE, Dowman LM, Hill CJ. Principles of rehabilitation and reactivation: interstitial lung disease, sarcoidosis and rheumatoid disease with respiratory involvement. Respiration. 2015;89(2):89–99.
3. Raghu G, Collard HR, Egan JJ, Martinez FJ, Behr J, Brown KK, et al. An official ATS/ERS/JRS/ALAT statement: idiopathic pulmonary fibrosis: evidence-based guidelines for diagnosis and management. Am J Respir Crit Care Med. 2011;183(6):788–824.

4. Olson AL, Swigris JJ, Lezotte DC, Norris JM, Wilson CG, Brown KK. Mortality from pulmonary fibrosis increased in the United States from 1992 to 2003. Am J Respir Crit Care Med. 2007;176(3):277–84.
5. Meltzer EB, Noble PW. Idiopathic pulmonary fibrosis. Orphanet J Rare Dis. 2008;3:8.
6. American Thoracic Society. Idiopathic pulmonary fibrosis: diagnosis and treatment. International consensus statement. American Thoracic Society (ATS), and the European Respiratory Society (ERS). Am J Respir Crit Care Med. 2000;161(2 Pt 1): 646–64.
7. Swigris JJ, Brown KK, Make BJ, Wamboldt FS. Pulmonary rehabilitation in idiopathic pulmonary fibrosis: a call for continued investigation. Respir Med. 2008;102(12):1675–80.
8. Swigris JJ, Kuschner WG, Jacobs SS, Wilson SR, Gould MK. Health-related quality of life in patients with idiopathic pulmonary fibrosis: a systematic review. Thorax. 2005;60(7): 588–94.
9. King TE Jr, Bradford WZ, Castro-Bernardini S, Fagan EA, Glaspole I, Glassberg MK, et al. A phase 3 trial of pirfenidone in patients with idiopathic pulmonary fibrosis. N Engl J Med. 2014;370(22):2083–92.
10. Noble PW, Albera C, Bradford WZ, Costabel U, Glassberg MK, Kardatzke D, et al. Pirfenidone in patients with idiopathic pulmonary fibrosis (CAPACITY): two randomised trials. Lancet. 2011;377(9779):1760–9.
11. Richeldi L, du Bois RM, Raghu G, Azuma A, Brown KK, Costabel U, et al. Efficacy and safety of nintedanib in idiopathic pulmonary fibrosis. N Engl J Med. 2014;370(22):2071–82.
12. Dowman L, Hill CJ, Holland AE. Pulmonary rehabilitation for interstitial lung disease. Cochrane Database Syst Rev. 2014;10:CD006322.
13. Ryerson CJ, Donesky D, Pantilat SZ, Collard HR. Dyspnea in idiopathic pulmonary fibrosis: a systematic review. J Pain Symptom Manag. 2012;43(4):771–82.
14. Mura M, Porretta MA, Bargagli E, Sergiacomi G, Zompatori M, Sverzellati N, et al. Predicting survival in newly diagnosed idiopathic pulmonary fibrosis: a 3-year prospective study. Eur Respir J. 2012;40(1):101–9.
15. Kozu R, Jenkins S, Senjyu H. Effect of disability level on response to pulmonary rehabilitation in patients with idiopathic pulmonary fibrosis. Respirology. 2011;16(8):1196–202.
16. Lama VN, Martinez FJ. Resting and exercise physiology in interstitial lung diseases. Clin Chest Med. 2004;25(3):435–53.. v
17. Holland AE. Exercise limitation in interstitial lung disease - mechanisms, significance and therapeutic options. Chron Respir Dis. 2010;7(2):101–11.
18. Spruit MA, Singh SJ, Garvey C, Zuwallack R, Nici L, Rochester C, et al. An official American thoracic society/European respiratory society statement: key concepts and advances in pulmonary rehabilitation. Am J Respir Crit Care Med. 2013;188(8):e13–64.
19. American Thoracic Society, American College of Chest Physicians. ATS/ACCP Statement on cardiopulmonary exercise testing. Am J Respir Crit Care Med. 2003;167(2):211–77.
20. Arena R, Sietsema KE. Cardiopulmonary exercise testing in the clinical evaluation of patients with heart and lung disease. Circulation. 2011;123(6):668–80.
21. Balady GJ, Arena R, Sietsema K, Myers J, Coke L, Fletcher GF, et al. Clinician's guide to cardiopulmonary exercise testing in adults: a scientific statement from the American Heart Association. Circulation. 2010;122(2):191–225.
22. Guazzi M, Adams V, Conraads V, Halle M, Mezzani A, Vanhees L, et al. EACPR/AHA scientific statement. Clinical recommendations for cardiopulmonary exercise testing data assessment in specific patient populations. Circulation. 2012;126(18):2261–74.
23. American College of Sports Medicine. ACSM's guidelines for exercise testing and prescription, vol. xxiv. 9th ed. Philadelphia, PA: Wolters Kluwer/Lippincott Williams & Wilkins Health; 2014. p. 456.
24. Fletcher GF, Ades PA, Kligfield P, Arena R, Balady GJ, Bittner VA, et al. Exercise standards for testing and training: a scientific statement from the American Heart Association. Circulation. 2013;128(8):873–934.
25. du Bois RM, Albera C, Bradford WZ, Costabel U, Leff JA, Noble PW, et al. 6-minute walk distance is an independent predictor of mortality in patients with idiopathic pulmonary fibrosis. Eur Respir J. 2014;43(5):1421–9.

26. Kawut SM, O'Shea MK, Bartels MN, Wilt JS, Sonett JR, Arcasoy SM. Exercise testing determines survival in patients with diffuse parenchymal lung disease evaluated for lung transplantation. Respir Med. 2005;99(11):1431–9.
27. Lederer DJ, Arcasoy SM, Wilt JS, D'Ovidio F, Sonett JR, Kawut SM. Six-minute-walk distance predicts waiting list survival in idiopathic pulmonary fibrosis. Am J Respir Crit Care Med. 2006;174(6):659–64.
28. Caminati A, Bianchi A, Cassandro R, Mirenda MR, Harari S. Walking distance on 6-MWT is a prognostic factor in idiopathic pulmonary fibrosis. Respir Med. 2009;103(1):117–23.
29. Lama VN, Flaherty KR, Toews GB, Colby TV, Travis WD, Long Q, et al. Prognostic value of desaturation during a 6-minute walk test in idiopathic interstitial pneumonia. Am J Respir Crit Care Med. 2003;168(9):1084–90.
30. Vainshelboim B, Kramer MR, Izhakian S, Lima RM, Oliveira J. Physical activity and exertional desaturation are associated with mortality in idiopathic pulmonary fibrosis. J Clin Med. 2016;5(8):E73.
31. Fell CD, Liu LX, Motika C, Kazerooni EA, Gross BH, Travis WD, et al. The prognostic value of cardiopulmonary exercise testing in idiopathic pulmonary fibrosis. Am J Respir Crit Care Med. 2009;179(5):402–7.
32. Triantafillidou C, Manali E, Lyberopoulos P, Kolilekas L, Kagouridis K, Gyftopoulos S, et al. The role of cardiopulmonary exercise test in IPF prognosis. Pulm Med. 2013;2013:514817.
33. Vainshelboim B, Oliveira J, Fox BD, Kramer MR. The prognostic role of ventilatory inefficiency and exercise capacity in idiopathic pulmonary fibrosis. Respir Care. 2016;61(8):1100–9.
34. Layton AM, Armstrong HF, Kim HP, Meza KS, D'Ovidio F, Arcasoy SM. Cardiopulmonary exercise factors predict survival in patients with advanced interstitial lung disease referred for lung transplantation. Respir Med. 2017;126:59–67.
35. Vainshelboim B, Oliveira J, Izhakian S, Unterman A, Kramer MR. Lifestyle behaviors and clinical outcomes in idiopathic pulmonary fibrosis. Respiration. 2018;95(1):27–34.
36. Vainshelboim B, Oliveira J, Fox BD, Adir Y, Ollech JE, Kramer MR. Physiological profile and limitations in exercise in idiopathic pulmonary fibrosis. J Cardiopulm Rehabil Prev. 2016;36(4):270–8.
37. ATS Committee on Proficiency Standards for Clinical Pulmonary Function Laboratories. ATS statement: guidelines for the six-minute walk test. Am J Respir Crit Care Med. 2002;166(1):111–7.
38. Holland AE, Spruit MA, Troosters T, Puhan MA, Pepin V, Saey D, et al. An official European Respiratory Society/American Thoracic Society technical standard: field walking tests in chronic respiratory disease. Eur Respir J. 2014;44(6):1428–46.
39. Singh SJ, Puhan MA, Andrianopoulos V, Hernandes NA, Mitchell KE, Hill CJ, et al. An official systematic review of the European Respiratory Society/American Thoracic Society: measurement properties of field walking tests in chronic respiratory disease. Eur Respir J. 2014;44(6):1447–78.
40. Vainshelboim B, Fox BD, Oliveira J, Kramer MR. Exercise training in idiopathic pulmonary fibrosis. Expert Rev Respir Med. 2016;10(1):69–77.
41. Tuppin MP, Paratz JD, Chang AT, Seale HE, Walsh JR, Kermeeen FD, et al. Predictive utility of the 6-minute walk distance on survival in patients awaiting lung transplantation. J Heart Lung Transplant. 2008;27(7):729–34.
42. Jastrzebski D, Gumola A, Gawlik R, Kozielski J. Dyspnea and quality of life in patients with pulmonary fibrosis after six weeks of respiratory rehabilitation. J Physiol Pharmacol. 2006;57(Suppl 4):139–48.
43. Ozalevli S, Karaali HK, Ilgin D, Ucan ES. Effect of home-based pulmonary rehabilitation in patients with idiopathic pulmonary fibrosis. Multidiscip Respir Med. 2010;5(1):31–7.
44. Swigris JJ, Fairclough DL, Morrison M, Make B, Kozora E, Brown KK, et al. Beneficial effects of pulmonary rehabilitation in idiopathic pulmonary fibrosis. Respir Care. 2011;56(6): 783–9.
45. Rammaert B, Leroy S, Cavestri B, Wallaert B, Grosbois JM. Home-based pulmonary rehabilitation in idiopathic pulmonary fibrosis. Rev Mal Respir. 2011;28(7):e52–7.

46. Ferreira A, Garvey C, Connors GL, Hilling L, Rigler J, Farrell S, et al. Pulmonary rehabilitation in interstitial lung disease: benefits and predictors of response. Chest. 2009;135(2):442–7.
47. Huppmann P, Sczepanski B, Boensch M, Winterkamp S, Schonheit-Kenn U, Neurohr C, et al. Effects of inpatient pulmonary rehabilitation in patients with interstitial lung disease. Eur Respir J. 2013;42(2):444–53.
48. Nishiyama O, Kondoh Y, Kimura T, Kato K, Kataoka K, Ogawa T, et al. Effects of pulmonary rehabilitation in patients with idiopathic pulmonary fibrosis. Respirology. 2008;13(3):394–9.
49. Holland AE, Hill CJ, Conron M, Munro P, McDonald CF. Short term improvement in exercise capacity and symptoms following exercise training in interstitial lung disease. Thorax. 2008;63(6):549–54.
50. Ryerson CJ, Cayou C, Topp F, Hilling L, Camp PG, Wilcox PG, et al. Pulmonary rehabilitation improves long-term outcomes in interstitial lung disease: a prospective cohort study. Respir Med. 2014;108(1):203–10.
51. Dowman LM, McDonald CF, Hill CJ, Lee AL, Barker K, Boote C, et al. The evidence of benefits of exercise training in interstitial lung disease: a randomised controlled trial. Thorax. 2017;72(7):610–9.
52. Tonelli R, Cocconcelli E, Lanini B, Romagnoli I, Florini F, Castaniere I, et al. Effectiveness of pulmonary rehabilitation in patients with interstitial lung disease of different etiology: a multicenter prospective study. BMC Pulm Med. 2017;17(1):130.
53. Arizono S, Taniguchi H, Sakamoto K, Kondoh Y, Kimura T, Kataoka K, et al. Endurance time is the most responsive exercise measurement in idiopathic pulmonary fibrosis. Respir Care. 2014;59(7):1108–15.
54. Vainshelboim B, Oliveira J, Yehoshua L, Weiss I, Fox BD, Fruchter O, et al. Exercise training-based pulmonary rehabilitation program is clinically beneficial for idiopathic pulmonary fibrosis. Respiration. 2014;88(5):378–88.
55. Jackson RM, Gomez-Marin OW, Ramos CF, Sol CM, Cohen MI, Gaunaurd IA, et al. Exercise limitation in IPF patients: a randomized trial of pulmonary rehabilitation. Lung. 2014;192:367.
56. Holland A, Hill C. Physical training for interstitial lung disease. Cochrane Database Syst Rev. 2008;4:CD006322.
57. du Bois RM, Weycker D, Albera C, Bradford WZ, Costabel U, Kartashov A, et al. 6-minute walk test in idiopathic pulmonary fibrosis: test validation and minimal clinically important difference. Am J Respir Crit Care Med. 2011;183(9):1231–7.
58. Garber CE, Blissmer B, Deschenes MR, Franklin BA, Lamonte MJ, Lee IM, et al. American College of Sports Medicine position stand. Quantity and quality of exercise for developing and maintaining cardiorespiratory, musculoskeletal, and neuromotor fitness in apparently healthy adults: guidance for prescribing exercise. Med Sci Sports Exerc. 2011;43(7):1334–59.
59. Chodzko-Zajko WJ, Proctor DN, Fiatarone Singh MA, Minson CT, Nigg CR, Salem GJ, et al. American College of Sports Medicine position stand. Exercise and physical activity for older adults. Med Sci Sports Exerc. 2009;41(7):1510–30.
60. World Health Organization. Global recommendations on physical activity for health. 2010. http://apps.who.int/iris/bitstream/10665/44399/1/9789241599979_eng.pdf.
61. Hallal PC, Andersen LB, Bull FC, Guthold R, Haskell W, Ekelund U, et al. Global physical activity levels: surveillance progress, pitfalls, and prospects. Lancet. 2012;380(9838):247–57.
62. Nakayama M, Bando M, Araki K, Sekine T, Kurosaki F, Sawata T, et al. Physical activity in patients with idiopathic pulmonary fibrosis. Respirology. 2015;20(4):640–6.
63. Wallaert B, Monge E, Le Rouzic O, Wemeau-Stervinou L, Salleron J, Grosbois JM. Physical activity in daily life of patients with fibrotic idiopathic interstitial pneumonia. Chest. 2013;144(5):1652–8.
64. Bahmer T, Kirsten AM, Waschki B, Rabe KF, Magnussen H, Kirsten D, et al. Prognosis and longitudinal changes of physical activity in idiopathic pulmonary fibrosis. BMC Pulm Med. 2017;17(1):104.
65. Vainshelboim B, Fox BD, Kramer MR, Izhakian S, Gershman E, Oliveira J. Short-term improvement in physical activity and body composition after supervised exercise training program in idiopathic pulmonary fibrosis. Arch Phys Med Rehabil. 2016;97(5):788–97.

66. Gaunaurd IA, Gomez-Marin OW, Ramos CF, Sol CM, Cohen MI, Cahalin LP, et al. Physical activity and quality of life improvements of patients with idiopathic pulmonary fibrosis completing a pulmonary rehabilitation program. Respir Care. 2014;59(12):1872–9.
67. Matthews CE, Chen KY, Freedson PS, Buchowski MS, Beech BM, Pate RR, et al. Amount of time spent in sedentary behaviors in the United States, 2003–2004. Am J Epidemiol. 2008;167(7):875–81.
68. Biswas A, Oh PI, Faulkner GE, Bajaj RR, Silver MA, Mitchell MS, et al. Sedentary time and its association with risk for disease incidence, mortality, and hospitalization in adults: a systematic review and meta-analysis. Ann Intern Med. 2015;162(2):123–32.
69. Bouchard C, Blair SN, Katzmarzyk PT. Less sitting, more physical activity, or higher fitness? Mayo Clin Proc. 2015;90(11):1533–40.
70. Atkins C, Baxter M, Jones A, Wilson A. Measuring sedentary behaviors in patients with idiopathic pulmonary fibrosis using wrist-worn accelerometers. Clin Respir J. 2018;12(2):746–53.
71. Leuchte HH, Mernitz P, Baezner C, Baumgartner RA, von Wulffen W, Neurohr C, et al. Self-report daily life activity as a prognostic marker of idiopathic pulmonary fibrosis. Respiration. 2015;90(6):460–7.
72. Kenney WL, Wilmore JH, Costill DL. Physiology of sport and exercise, vol. xvii. 5th ed. Champaign, IL: Human Kinetics; 2012. p. 621.
73. Ross R, Blair SN, Arena R, Church TS, Despres JP, Franklin BA, et al. Importance of assessing cardiorespiratory fitness in clinical practice: a case for fitness as a clinical vital sign: a scientific statement from the American Heart Association. Circulation. 2016;134(24):e653–e99.
74. Kenn K, Gloeckl R, Behr J. Pulmonary rehabilitation in patients with idiopathic pulmonary fibrosis—a review. Respiration. 2013;86(2):89–99.
75. Manali ED, Lyberopoulos P, Triantafillidou C, Kolilekas LF, Sotiropoulou C, Milic-Emili J, et al. MRC chronic dyspnea scale: relationships with cardiopulmonary exercise testing and 6-minute walk test in idiopathic pulmonary fibrosis patients: a prospective study. BMC Pulm Med. 2010;10:32.
76. Keyser RE, Woolstenhulme JG, Chin LM, Nathan SD, Weir NA, Connors G, et al. Cardiorespiratory function before and after aerobic exercise training in patients with interstitial lung disease. J Cardiopulm Rehabil Prev. 2015;35(1):47–55.
77. Vainshelboim B, Kramer MR, Fox BD, Izhakian S, Sagie A, Oliveira J. Supervised exercise training improves exercise cardiovascular function in idiopathic pulmonary fibrosis. Eur J Phys Rehabil Med. 2017;53(2):209–18.
78. Nathan SD, Basavaraj A, Reichner C, Shlobin OA, Ahmad S, Kiernan J, et al. Prevalence and impact of coronary artery disease in idiopathic pulmonary fibrosis. Respir Med. 2010;104(7):1035–41.
79. Nakazawa A, Cox NS, Holland AE. Current best practice in rehabilitation in interstitial lung disease. Ther Adv Respir Dis. 2017;11(2):115–28.
80. Garvey C. Interstitial lung disease and pulmonary rehabilitation. J Cardiopulm Rehabil Prev. 2010;30(3):141–6.
81. Celli BR, MacNee W, Force AET. Standards for the diagnosis and treatment of patients with COPD: a summary of the ATS/ERS position paper. Eur Respir J. 2004;23(6):932–46.
82. Kozu R, Senjyu H, Jenkins SC, Mukae H, Sakamoto N, Kohno S. Differences in response to pulmonary rehabilitation in idiopathic pulmonary fibrosis and chronic obstructive pulmonary disease. Respiration. 2011;81(3):196–205.
83. Holland AE, Dowman L, Fiore J Jr, Brazzale D, Hill CJ, McDonald CF. Cardiorespiratory responses to 6-minute walk test in interstitial lung disease: not always a submaximal test. BMC Pulm Med. 2014;14:136.
84. Ries AL, Bauldoff GS, Carlin BW, Casaburi R, Emery CF, Mahler DA, et al. Pulmonary rehabilitation: joint ACCP/AACVPR evidence-based clinical practice guidelines. Chest. 2007;131(5 Suppl):4S–42S.
85. Nici L, Donner C, Wouters E, Zuwallack R, Ambrosino N, Bourbeau J, et al. American Thoracic Society/European Respiratory Society statement on pulmonary rehabilitation. Am J Respir Crit Care Med. 2006;173(12):1390–413.

Exercise in Pulmonary Vascular Diseases

8

Pierantonio Laveneziana and Louis Laviolette

Abstract

Pulmonary arterial hypertension and chronic thromboembolic pulmonary hypertension are the most common diseases of pulmonary vasculature. The physiological derangements of pulmonary hypertension result in characteristic abnormalities observed during dynamic exercise and often lead to dyspnoea and exercise intolerance. Impaired cardiac function results in reduced aerobic capacity, low anaerobic threshold and reduced value of the relationship between oxygen uptake and work rate ($\Delta V'O_2/\Delta WR$). Both high physiologic dead space and chemosensitivity contribute to elevated ratio of minute ventilation to CO_2 output ($V'_E/V'CO_2$) during exercise testing. Consequently, resting hypocapnia with low end-tidal PCO_2 throughout exercise is typically observed and is related to the severity of disease. Exertional hypoxaemia is also a variable but frequent finding during exercise, which can be related to ventilation-perfusion heterogeneity, low mixed venous O_2 content from impaired cardiac output and right-to-left shunting through a patent foramen ovale. Even in the absence of significant resting airflow obstruction, dynamic hyperinflation can occur in pulmonary vascular diseases, which contributes to exertional dyspnoea and exercise intolerance. Peripheral muscle dysfunction is another common component of exercise pathophysiology in these conditions.

P. Laveneziana (✉)
Sorbonne Université, INSERM, UMRS1158 Neurophysiologie respiratoire expérimentale et clinique, Paris, France

AP-HP, Groupe Hospitalier Pitié-Salpêtrière Charles Foix, Service des Explorations Fonctionnelles de la Respiration, de l'Exercice et de la Dyspnée du Département "R3S", Paris, France
e-mail: pierantonio.laveneziana@aphp.fr

L. Laviolette
Faculté de médecine, Université Laval, Québec, QC, Canada

Centre de recherche de l'Institut Universitaire de Cardiologie et de Pneumologie de Québec (CRIUCPQ), Québec, QC, Canada

A. Cogo et al. (eds.), *Exercise and Sports Pulmonology*,
https://doi.org/10.1007/978-3-030-05258-4_8

8.1 Introduction

Pulmonary hypertension is defined as a resting mean pulmonary arterial pressure (mPAP) ≥25 mmHg, which may result from primary diseases of the pulmonary vasculature, left heart disease, lung disease and systemic diseases [1]. Pulmonary arterial hypertension (PAH) and chronic thromboembolic pulmonary hypertension (CTEPH) are primary diseases of the pulmonary vasculature caused by obstruction, inflammation and remodelling of the pulmonary arteries and arterioles, endothelial dysfunction, vasoconstriction and thrombosis [2]. PAH may be idiopathic or caused by underlying connective tissue diseases, congenital heart disease, genetic mutations, drugs and toxins, portal hypertension or infection with human immunodeficiency virus or schistosomiasis [1]. CTEPH is a rare complication of pulmonary thromboembolism in which there is persistent obstruction of large- and medium-sized pulmonary arteries with remodelling of distal small vessels and progressive pulmonary hypertension [3, 4]. Over time, patients with pulmonary vascular diseases develop progressive increases in mPAP and pulmonary vascular resistance, which ultimately leads to right heart failure and death.

Cardiopulmonary exercise testing (CPET) is very sensitive in detecting possible impairments in a patient with early pulmonary vascular disease. Beside revealing common and non-specific symptoms like dyspnoea and exercise intolerance, CPET can highlight abnormal exercise response patterns suggestive of pulmonary vascular disease in patients with undifferentiated dyspnoea [5]. In addition to functional assessment, the CPET is helpful for the evaluation of responses to treatment and estimate prognosis [6].

This chapter will focus primarily on the dynamic exercise pathophysiology and patterns of exercise responses during CPET in patients with pulmonary hypertension and increased pulmonary vascular resistance, without significant left heart disease (Group 2 pulmonary hypertension) or a significant obstructive or restrictive ventilatory defect (Group 3 pulmonary hypertension). The features and impact of pulmonary hypertension secondary to other lung and heart diseases will not be discussed in this chapter.

8.2 Exercise Pathophysiology in PAH and CTEPH: General Hallmarks

From a pathophysiological point of view, PAH and CTEPH are characterised by obliteration and consequent obstruction of pulmonary arteries, vascular inflammation and consequent remodelling and endothelial dysfunction, which all give rise to increased pulmonary arterial resistance and elevated pulmonary arterial pressure [6]. The consequence of all this is that dead space (V_D/V_T) ventilation increases because of the reduced perfusion of well-ventilated alveoli, which is reflected "mainly" as a high ratio of minute ventilation (V'_E) to CO_2 output ($V'CO_2$) and expressed as $V'_E/V'CO_2$. During exercise, cardiac output (CO) must increase to match oxygen transport to the increasing demand by locomotor muscles [6].

Increasing pulmonary blood flow during exercise is normally guaranteed by vascular distension and recruitment to keep resistance low for the right ventricle (RV). However, in patients with pulmonary vascular disease, fixed vascular remodelling hinders normal recruitment and distension and translates into a persistently high vascular resistance, and therefore increases in CO during exercise give rise to further increases in mPAP. This progressive elevation in afterload curtails the ability of the RV to increase stroke volume, and therefore increases in CO during exercise strongly rely on heart rate (HR) [6].

High RV pressure and RV dilation lead to interventricular septal shift, which, along with reduced pulmonary venous return to the left atrium, limits left ventricular (LV) diastolic filling, systemic CO and tissue oxygen transport [6]. Arterial desaturation may also occur during exercise due to a combination of low mixed venous oxygen saturation, relative low alveolar-capillary diffusing capacity, high physiologic dead space or right-to-left shunting through a patent foramen ovale. Hypoxaemia further worsens the blunted tissue oxygen delivery, being conducive to the early onset of lactic acidosis and reduced anaerobic threshold (AT) that develop in the presence of a reduced CO [6]. Furthermore, hypoxaemia, lactic acidosis and higher VCO_2 stemming from anaerobic metabolism all contribute to an excessive increase in V'_E during exercise. These pathophysiologic mechanisms result in characteristic pattern of abnormalities observed during CPET in patients with pulmonary vascular diseases (Table 8.1) and depict the various potential contributors to dyspnoea, leg fatigue and exercise intolerance (Fig. 8.1).

8.3 Cardiovascular Abnormalities

During cardiac systole, both the systemic and pulmonary circulations must handle the same volume of blood. The latter, however, is normally at 10% of the former's pressure. During dynamic exercise, cardiac output (CO) must increase to match oxygen delivery to demand by peripheral muscles. Even the greatly increased cardiac output of exercising healthy subjects will lead to only a modest increase in mPAP pressure, mainly because of the large capacitance of the pulmonary circulation [7], which is the consequence of the increase in left atrial pressure.

Even in early pulmonary vascular disease, when resting mPAP is not yet elevated, there is a loss of vascular distensibility [8], and mPAP rises disproportionately to CO [9, 10]. In severe PAH and CTEPH, the pulmonary vasculature cannot accommodate increased pulmonary blood flow, resulting in further and excessive increases in the right ventricular (RV) afterload during exercise [11]. Because of ventricular interdependence, severe RV pressure overload shifts the interventricular septum to the left during diastole and impairs left ventricular filling [12–16], limiting maximal CO and oxygen delivery. Thus, in patients with pulmonary vascular diseases, changes in CO during exercise are mostly mediated by increasing heart rate (HR) rather than increasing stroke volume [15, 17, 18]. The ability to increase CO during exercise is a more important determinant of peak exercise capacity than the resting CO in patients with PAH and CTEPH, as it reflects the severity of the

Table 8.1 Typical CPET abnormalities in patients with pulmonary vascular diseases

	PAH	CTEPH	PVOD
Metabolic and cardiovascular			
Peak $V'O_2$	↓	↓	↓
$V'O_2$ at AT	↓	↓	↓↓
$V'O_2$/WR	↓	↓	↓
Peak O_2 pulse	↓	↓	↓
Ventilation and mechanics			
Peak V'_E	↓	↓	↓
Breathing reserve	Normal	Normal	Normal
Dynamic hyperinflation	Possible	Possible	?
Gas exchange			
$V'_E/V'CO_2$ slope	↑	↑↑	↑↑
$V'_E/V'CO_2$ at AT	↑	↑↑	↑↑
OUE	↓	↓	?
OUES	↓	↓	?
OUEP	↓	↓↓	?
$P_{ET}CO_2$ (peak and at AT)	↓	↓↓	↓↓
SaO_2	↓	↓↓	↓↓
Peak $P_{a\text{-}ET}CO_2$	↑	↑↑	↑↑
Peak $P_{A\text{-}a}O_2$	↑	↑↑	↑↑
Peak V_D/V_T	↑	↑↑	↑↑

CPET cardiopulmonary exercise testing, *PAH* pulmonary arterial hypertension, *CTEPH* chronic thromboembolic pulmonary hypertension, *PVOD* pulmonary veno-occlusive disease, $V'O_2$ oxygen consumption, *AT* anaerobic threshold, *WR* work rate, *O_2 pulse* peak $V'O_2$-to-heart rate ratio at peak exercise, V'_E minute ventilation, $V'_E/V'CO_2$ ratio of minute ventilation to carbon dioxide production ($V'CO_2$), *OUE* oxygen uptake efficiency ($V'O_2/V'CO_2$), *OUES* oxygen uptake efficiency slope, *OUEP* oxygen uptake efficiency plateau, *$PETCO_2$* end-tidal pressure of carbon dioxide, *SaO_2* arterial oxygen saturation, $P_{A\text{-}a}O_2$ alveolar-arterial oxygen pressure gradient at peak exercise, $P_{a\text{-}ET}CO_2$ arterial to end-tidal carbon dioxide pressure gradient at peak exercise, V_D/V_T physiologic dead space fraction as ratio of dead space (V_D) to tidal volume (V_T) at peak exercise

underlying pulmonary vascular disease and the ability of the RV to adapt to it [19]. In patients with chronic thromboembolism but without manifest pulmonary hypertension at rest, the mPAP/CO slope during exercise is abnormally high and stroke volume increases minimally, indicating that RV stroke volume response is impaired early in the disease course [20].

These pathophysiological adaptations to the increased pulmonary arterial pressures all contribute to the cardiovascular limitation to exercise during CPET. Because of high RV afterload and low maximal CO, oxygen delivery to the skeletal muscle is impaired, manifesting as reduced aerobic capacity (low peak $V'O_2$) and early shift to anaerobic metabolism for a given $V'O_2$ (i.e., low anaerobic threshold) [21]. While maximal work rate (WR) and maximal $V'O_2$ are often reduced, the $\Delta V'O_2/\Delta WR$ relationship is also low (<8–9 $mL \cdot min^{-1}\ W^{-1}$) compared to healthy individuals or patients with left ventricular failure, reflecting impaired CO and/or abnormal peripheral muscle O_2 utilisation (Fig. 8.2) [21–25].

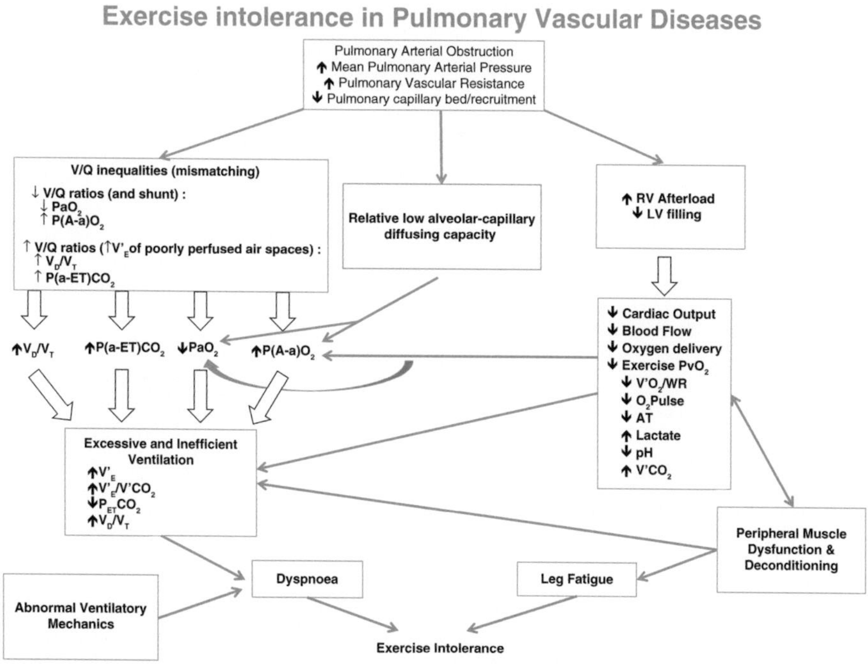

Fig. 8.1 Pathophysiology and mechanisms of exercise intolerance in pulmonary hypertension. Pulmonary vascular obstruction results in high ventilation-to-perfusion ratios and impaired cardiac output and can result in hypoxaemia due to right-to-left shunting through a patent foramen ovale. Inefficient ventilation proposes high ventilatory demand, high $V'_E/V'CO_2$ and V_D/V_T and low $P_{ET}CO_2$. Cardiac limitation and peripheral muscle abnormalities result in a low anaerobic threshold, early-onset lactic acidosis and increased $V'CO_2$, which provide further stimulation for excessive ventilation. Ventilatory mechanical constraints on tidal volume expansion also contribute to dyspnoea during exercise. *Abbreviations*: *V′/Q′* ventilation-to-perfusion ratio, *RV* right ventricle, *LV* left ventricle, V'_E minute ventilation, $V'_E/V'CO_2$ ratio of minute ventilation to carbon dioxide production, $P_{ET}CO_2$ end-tidal pressure of carbon dioxide, V_D/V_T dead space to tidal volume fraction, $V'O_2$ oxygen consumption, *WR* work rate, *O_2 pulse* $V'O_2$-to-heart rate ratio, $V'CO_2$ carbon dioxide production, PvO_2 venous partial pressure of oxygen, PaO_2 arterial partial pressure of oxygen, $PaCO_2$ arterial partial pressure of carbon dioxide. This is an original figure, no permission is required

Stroke volume is an important contributor to the increase in CO during exercise, and we have previously discussed that it is significantly impaired in pulmonary vascular disease. It is, however, very difficult to measure and a surrogate measure can be used.

Given Fick's equation where CO is equal to $V'O_2$ divided by the arteriovenous O_2 difference (CaO_2–CvO_2) and that CO = HR × stroke volume, the equation can be rearranged as $V'O_2$/HR (O_2 pulse) = stroke volume × (CaO_2–CvO_2). Thus, in the absence of arterial desaturation, a reduced O_2 pulse reflects an impaired stroke volume response during exercise. This means that cardiac output solely depends on increasing heart rate, leading to a decreased and flattened profile of

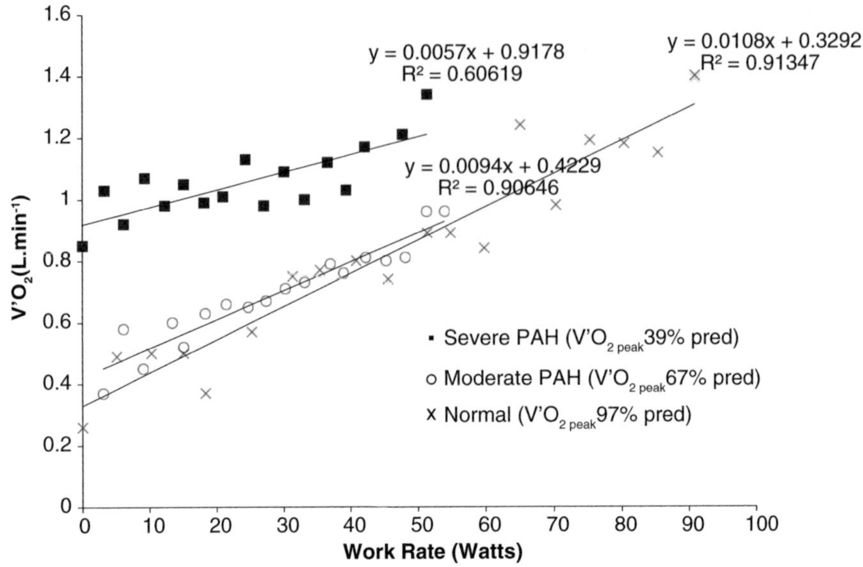

Fig. 8.2 Comparison of oxygen consumption ($V'O_2$) to work rate (WR) relationships. A normal individual with peak $V'O_2$ of 97% has a $V'O_2$/WR slope of 10.8 mL per Watt. A patient with pulmonary arterial hypertension (PAH, pulmonary vascular resistance 9.3 Wood units) and a preserved cardiac index (CI = 2.7 $L{\cdot}min{\cdot}m^{-2}$) and moderately reduced peak $V'O_2$ of 67% predicted has a borderline reduction in the $V'O_2$/WR slope of 9.4 mL per Watt. The patient with PAH and severe reduction in peak $V'O_2$ (39% predicted) demonstrates a reduced $V'O_2$/WR slope of 5.7 mL per Watt. Note that the difference in y-intercept ($V'O_2$) at WR of 0 Watts is largely related to variability in body mass between these individuals. Original figure. Data from authors' own laboratory

the O_2 pulse ($V'O_2$/HR) [21, 23]. Poor RV function and stroke volume response may lead to low systolic blood pressure (SBP) during exercise, and symptoms of pre-syncope or even syncope may occur. A peak exercise SBP < 120 mmHg during CPET should be considered an ominous sign [26]. Vagal reactivation after exercise is an important mechanism underlying HR recovery in the first 30 s–60 s after exercise and is abnormally slow in individuals with cardiac impairment [27]. Recovery of HR after exercise is delayed in PAH patients compared to controls, and slower HR recovery (<18 beat per minute decrease in the first minute post-exercise) is associated with worse resting haemodynamics, lower peak $V'O_2$ and a worse prognosis [28, 29].

Thus, the primary abnormalities of cardiovascular variables during exercise testing in patients with moderate to severe pulmonary vascular disease are (1) reduced peak $V'O_2$ and peak WR, (2) low anaerobic threshold, (3) reduced $\Delta V'O_2/\Delta WR$, (4) low and flattened O_2 pulse and (5) low maximal HR with delayed HR recovery (Fig. 8.1 and Table 8.1).

8.3.1 Ventilatory Abnormalities

For most patients with pulmonary vascular disease, exercise is not limited by encroachment upon their predicted maximal ventilatory capacity; ventilation (V'_E) at peak exercise is usually low [21, 22, 30, 31]. In the absence of concurrent asthma or chronic obstructive pulmonary disease, mechanical ventilatory constraint (dynamic hyperinflation) is not expected in pulmonary vascular diseases during CPET and, if present, is not as clinically relevant as in COPD.

Resting spirometry in patients with PAH is usually normal or may show mild restriction [32] or reduced mean expiratory flow (MEF) at 75%, 50% and 25% of vital capacity and increased residual volume-to-total lung capacity ratio (RV/TLC), suggestive of peripheral airways obstruction and gas trapping [33, 34].

Breathing patterns during exercise are more rapid and shallow in patients with PAH as opposed to normal individuals. Compared to healthy controls, up to 60% of PAH patients exhibit a reduction in inspiratory capacity during exercise, suggesting dynamic hyperinflation or impaired inspiratory muscle function [33, 35]. Even in the setting of a normal resting FEV_1/FVC, the presence of expiratory flow limitation and rapid shallow breathing patterns during exercise can promote dynamic hyperinflation in some PAH patients, which may lead to more severe dyspnoea (Figs. 8.3 and 8.4) [33].

Diaphragmatic muscle atrophy and weakness are present in patients with severe PAH or CTEPH [36–38] but did not appear to be involved in the dynamic reduction in

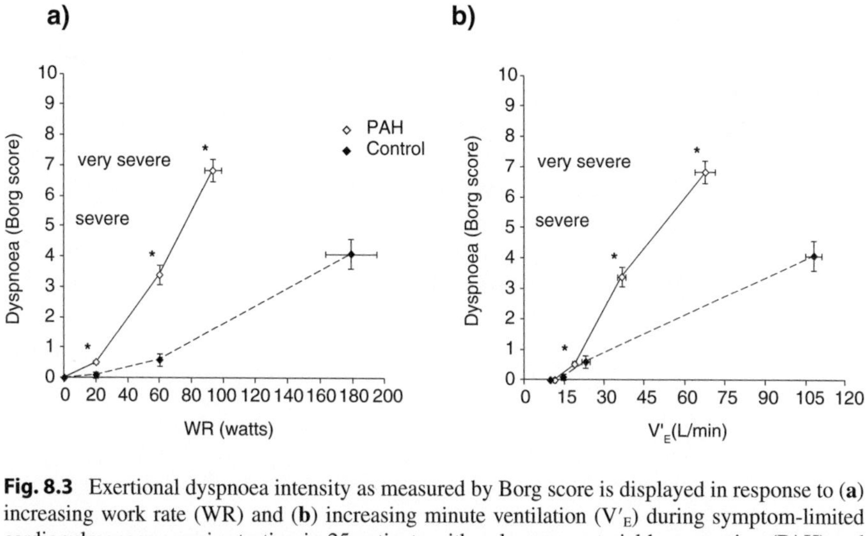

Fig. 8.3 Exertional dyspnoea intensity as measured by Borg score is displayed in response to (**a**) increasing work rate (WR) and (**b**) increasing minute ventilation (V'_E) during symptom-limited cardiopulmonary exercise testing in 25 patients with pulmonary arterial hypertension (PAH) and 10 healthy control subjects. *: $p < 0.05$, PAH vs. healthy control at rest and standardised exercise work rates (20–60 W) and peak exercise. From reference [33], with permission

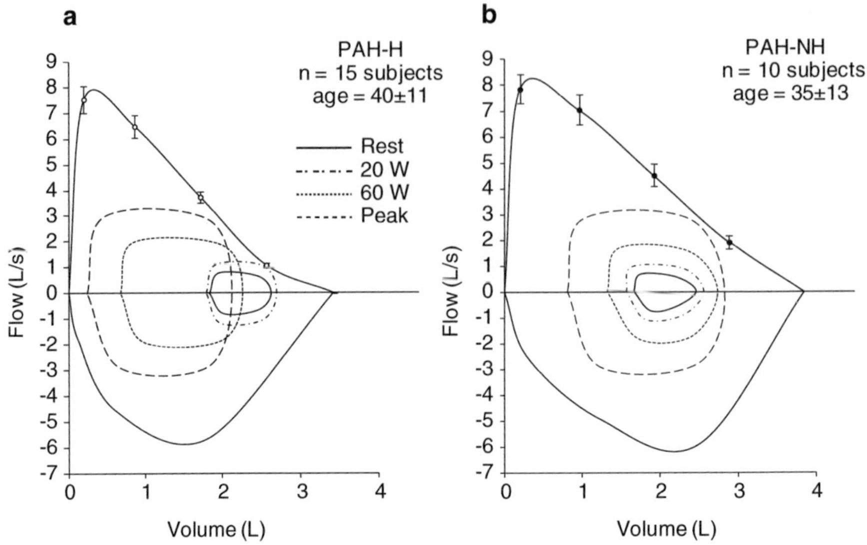

Fig. 8.4 Maximal and tidal flow-volume loops (average data) are shown at rest and during incremental cycle exercise in patients with pulmonary arterial hypertension (PAH) (**a**) with hyperinflation (PAH-H; n = 15, age 40 ± 11 years) and (**b**) without hyperinflation (PAH-NH; n = 10, age 35 ± 13 years). Tidal flow-volume loops are provided at rest, early in exercise (at 20 W), late in exercise (at 60 W) and at peak exercise. Note a significant decrease in dynamic inspiratory capacity during exercise in PAH-H compared with PAH-NH. From reference [33], with permission

IC when oesophageal manometry was performed during CPET [39]. Whether interventions such as supplemental oxygen or bronchodilators reduce or delay the onset of dynamic hyperinflation in this disease remains to be determined in these patients.

The efficiency of ventilation can be illustrated with the relationship between V'_E and carbon dioxide output ($V'CO_2$): less ventilation will be required to eliminate CO_2 in a more efficient system. It is usually reported as the $V'_E/V'CO_2$ slope or the lowest (or "nadir") value of $V'_E/V'CO_2$ during exercise. The $V'_E/V'CO_2$ slope is determined by the arterial PCO_2 ($PaCO_2$) and the physiologic dead space (V_D/V_T) according to Eq. (1):

$$V'_E / V'CO_2 = \frac{863}{PaCO_2 \left(1 - V_D / V_T\right)} \tag{1}$$

The V_D/V_T is calculated from the Enghoff modification to the Bohr equation in Eq. (2):

$$V_D / V_T = \frac{PaCO_2 - P_{\bar{E}}CO_2}{PaCO_2} \tag{2}$$

where $P_{\bar{E}}CO_2$ is the mixed expired breath PCO_2.

Ventilatory inefficiency (i.e. high $V'_E/V'CO_2$) and gas exchange abnormalities are hallmark features of pulmonary vascular diseases (Fig. 8.5a). In normal individuals <60 years old, the 95% confidence interval upper limit for $V'_E/V'CO_2$ slope is 33 and the $V'_E/V'CO_2$ nadir is 34 [40]. In PAH patients, $V'_E/V'CO_2$ slope and nadir are usually significantly increased compared to normal

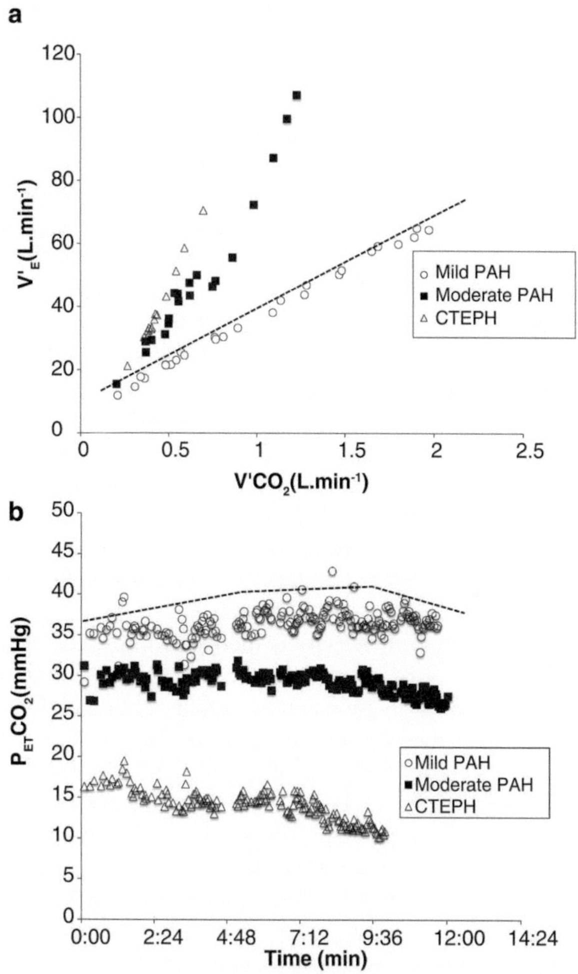

Fig. 8.5 (**a**) Ventilation (V'_E) plotted against CO_2 output ($V'CO_2$) for patients with mild (circles) pulmonary arterial hypertension (PAH), moderate PAH (solid squares) and chronic thromboembolic pulmonary hypertension (CTEPH, triangles). The dashed line represents the upper limit of normal. Note that in the patient with mild PAH, the $V'_E/V'CO_2$ slope is only mildly abnormal ($V'_E/V'CO_2$ slope = 30), whereas the patient with moderate PAH and severe CTEPH has significantly elevated $V'_E/V'CO_2$ slopes of 84–115, respectively. (**b**) End-tidal PCO_2 ($P_{ET}CO_2$) plotted against time for the same patients in (**a**). Note that the patient with mild PAH (circles) exhibits a slight increase in $P_{ET}CO_2$ during early exercise, similar to the predicted normal response (dashed line). The patient with moderate PAH exhibits a flat $P_{ET}CO_2$ during early exercise with a terminal decline coinciding with hyperventilation after the anaerobic threshold. The patient with severe CTEPH demonstrates a progressive decrease in $P_{ET}CO_2$ characteristic of severe pulmonary vascular disease. Original Fig. (**a** and **b**). Data from authors' own laboratory

individuals and are even higher than in patients with left ventricular failure, despite a similar degree of exercise impairment [23, 30, 41–43]. In patients with chronic thromboembolism without pulmonary hypertension, the $V'_E/V'CO_2$ slope, $V'_E/V'CO_2$ at anaerobic threshold and V_D/V_T are higher than in controls, indicating that ventilatory inefficiency can result from vascular obstruction and ventilation-perfusion (V'/Q') inequality, even before overt pulmonary hypertension and impaired RV function develop [44, 45]. Compared to PAH, CTEPH patients have even greater $V'_E/V'CO_2$ slope and nadir values and higher V_D/V_T at peak exercise [46, 47].

In PAH, the severity of increase in $V'_E/V'CO_2$ is related to the degree of elevation in mPAP [42, 48] and is a major determinant of peak VO_2 and New York Heart Association functional class [21]. Some authors have suggested that a combination of findings of low peak VO_2 and low anaerobic threshold, with preserved breathing reserve and $V'_E/V'CO_2$ at the anaerobic threshold >34, has 88% specificity and 85% accuracy for pulmonary vascular limitation to exercise [49].

High $V'_E/V'CO_2$ reflects wasted ventilation and is usually attributed to high V_D/V_T from ventilation-perfusion inequality, but it can also be related to high chemosensitivity from sympathetic nervous system hyperactivity or a low $PaCO_2$ set-point [50–53]. Although resting and peak exercise V_D/V_T are elevated in pulmonary vascular diseases, resting hypocapnia and exercise hyperventilation are common observations that correlate with disease severity, supporting the important contributions of chemoreceptor and sympathetic neural input and the $PaCO_2$ set-point to ventilatory inefficiency [31, 41, 46–48, 50, 52, 54–56]. The V_D/V_T calculated from Eq. (2) is sensitive to high levels of ventilation and to rapid-shallow breathing patterns; therefore high V_D/V_T may also reflect increased chemosensitivity [50–53].

Sympathetic nervous system activity is increased in patients with pulmonary hypertension and is a factor of decreased exercise capacity and worse prognosis [51, 57, 58]. The enhanced chemoreceptor output stimulates hyperventilation, which can be driven by hypoxaemia, low cardiac output and neural afferents from metabolic ergoreceptors in the peripheral muscles [50, 59–61]. Local accumulation of H^+ in skeletal muscles contributes to ergoreceptor-mediated stimulation of ventilation in patients with impaired cardiac function [62]. Right ventricular or right atrial distension may also mediate hyperventilation through sympathetic neural reflexes [56, 63]. This explains why the $V'_E/V'CO_2$, an integrated variable reflecting not only gas exchange, but cardiovascular and autonomic nervous system dysfunction, is associated with clinical outcomes in PAH and CTEPH [64, 65].

8.3.2 Gas Exchange Abnormalities

As a consequence of alveolar hyperventilation, the end-tidal PCO_2 ($P_{ET}CO_2$) measured at the mouth during CPET is frequently decreased in patients with PAH or CTEPH and does not exhibit a normal pattern of gradual increase between rest and the anaerobic threshold, as it remains constant or decreases further (Fig. 8.5b) [22, 47, 48, 66]. A $P_{ET}CO_2$ of <40 mmHg at the anaerobic threshold may suggest underlying pulmonary vascular disease, whereas $P_{ET}CO_2 < 20$ mmHg is unusual in other diseases and raises strong suspicion of pulmonary vascular disease in a patient with dyspnoea of unknown aetiology [48, 67, 68]. Patients with CTEPH or pulmonary veno-occlusive disease (PVOD), another rare pulmonary vascular disease, tend to have even lower resting and peak exercise $P_{ET}CO_2$ values than patients with idiopathic PAH (Table 8.1) [47, 66, 68].

Arterial oxygen desaturation and wide alveolar-arterial O_2 ($P_{A\text{-}a}O_2$) gradient are common (Fig. 8.6) but not universally observed in patients with pulmonary vascular diseases [31, 47, 55, 60, 68]. In contrast, significant arterial desaturation and hypoxaemia ($PaO_2 < 60$ mmHg) are much more frequent, while also being rarely observed in patients with heart failure. Therefore, desaturation may suggest the presence of underlying pulmonary vascular disease when present in an undifferentiated dyspnoeic patient [69–71]. The widened $P_{A\text{-}a}O_2$ and arterial desaturation are primarily

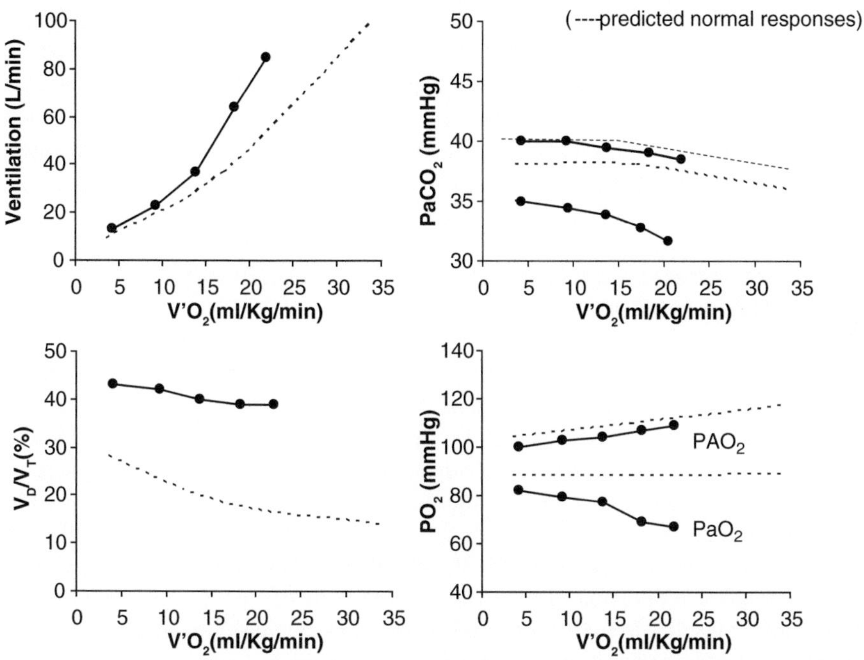

Fig. 8.6 Gas exchange abnormalities for a patient with pulmonary arterial hypertension (PAH). Dashed lines represent predicted normal responses. Note the excessive ventilation for a given $V'O_2$ (upper left panel) with high physiologic dead space (V_D/V_T) at rest and throughout exercise (lower left panel). The arterial CO_2 pressure ($PaCO_2$) is low at rest and decreases early during exercise to a greater extent than in a normal individual (upper right panel). The arterial oxygen pressure (PaO_2) decreases abnormally during exercise despite a normal increase in alveolar oxygen pressure (PAO_2), resulting in a wide and increasing alveolar-arterial O_2 difference at peak exercise (lower right panel). Original figure. Data from authors' own laboratory

related to low mixed venous PO_2 returning to the pulmonary circulation as a consequence of impaired cardiac function and oxygen delivery to peripheral muscles, which is exacerbated by ventilation-perfusion inequality in the lung [72, 73]. A pattern of exercise-induced hypoxaemia preceded by a sudden and sustained decrease in $P_{ET}CO_2$ and increase in end-tidal PO_2 ($P_{ET}O_2$) and the $V'_E/V'CO_2$ in patients with pulmonary vascular disease suggests the development of a right-to-left shunt through a patent foramen ovale (PFO) [60, 74]. This occurs when right atrial pressure rises high enough during exercise to open the PFO, shunting hypoxaemic and acidaemic blood to the systemic circulation, which acutely stimulates peripheral chemoreceptors and hyperventilation.

Normally, $P_{ET}CO_2$ increases during exercise in healthy individuals as a result of larger tidal volume and higher PCO_2 in venous blood returning to the lungs (Fig. 8.5b). Since $P_{ET}CO_2$ rises and $PaCO_2$ remains stable (or even decreases slightly) during exercise, the difference between $PaCO_2$ and $P_{ET}CO_2$ ($P_{(a\text{-}ET)}CO_2$) is slightly positive at rest and becomes negative in most normal individuals [75, 76]. In patients with pulmonary vascular disease, the excessive and inefficient V'_E driven by chemoreceptor stimulation often leads to very low $P_{ET}CO_2$ near peak exercise. Meanwhile, because of high physiologic dead space, ventilation-perfusion inequalities and rapid shallow breathing patterns, the arterial PCO_2 does not change markedly, leading to a positive $P_{(a\text{-}ET)}CO_2$ at rest and exercise [47, 55, 66, 68]. Therefore, a positive value for $P_{(a\text{-}ET)}CO_2$ from arterial blood gases performed at peak exercise reflects impaired gas exchange and/or augmented chemoreflexes.

8.3.3 Peripheral Muscle Function and Exercise in Pulmonary Hypertension

Deconditioning and peripheral muscle abnormalities are important contributors to exercise intolerance. In congestive heart failure, which shares similar limitations in cardiac output reserve as PAH and CTEPH, oxygen transport and diffusion at the level of the skeletal muscle are abnormal [77]. However, tissue oxygen saturation, oxygen extraction and muscle microcirculatory function may be impaired to an even greater degree in PAH compared with left heart failure [78, 79]. The peripheral muscle in PAH patients is structurally and functionally abnormal, with a lower relative proportion of type I fibres and reduced quadriceps, forearm and respiratory muscle strength compared to controls, which may be an important determinant of low peak VO_2 [25, 80]. Respiratory muscle strength has also been shown to be about 40% lower in CTEPH patients [37]. The mechanism of generalised skeletal muscle dysfunction in PAH may be a result of microcirculation rarefaction and an imbalance in angiogenic factors [24]. Improvements in exercise capacity with exercise training in individuals with heart failure or peripheral vascular disease [81] have been linked to improvements in skeletal muscle microcirculatory density, capillary-to-fibre ratio and mitochondrial volume [82], which may be mechanisms by which training can improve exercise capacity in stable patients with PAH [83, 84].

8.3.4 Prognostic Utility of Cardiopulmonary Exercise Testing

Several studies have shown that CPET variables independently predict prognosis in PAH and CTEPH patients. PAH patients with a peak $V'O_2$ less than 11 $mL \cdot min^{-1} \cdot kg^{-1}$ or a $V'_E/V'CO_2$ slope ≥ 45 are considered at high risk with an estimated 1-year mortality of >10% according to the European Society of Cardiology/European Respiratory Society guidelines [1]. Therefore, potential treatment targets for PAH patients have been established at obtaining peak $V'O_2 > 15$ $mL \cdot min^{-1} \cdot kg^{-1}$ or > 65% predicted and a $V'_E/V'CO_2$ slope of <36 [1, 85].

Peak $V'O_2$ and $V'_E/V'CO_2$ have been associated with survival in several studies comprising PAH and CTEPH patients [26, 64, 65, 86]. Wensel and colleagues demonstrated that peak $V'O_2$ provides additional prognostic value to resting haemodynamics in patients with PAH [87]. Those with a low $V'O_2$ (<46.3% predicted) and pulmonary vascular resistance (PVR) > 16 Wood units had a particularly dire prognosis, while patients with peak $V'O_2 \geq 46.3\%$ predicted and a PVR < 11.6 Wood units had >90% 5-year survival.

Echocardiographic assessment of RV function in conjunction with CPET may provide incremental prognostic utility. Badagliacca and colleagues found that resting RV fractional area change on echocardiogram, in combination with the O_2 pulse from CPET (which reflect RV function and stroke volume), was an independent predictor of outcome in patients with idiopathic PAH [88]. Patients with RV fractional area change >26.5% and a peak O_2 pulse >8.0 $mL \cdot beat^{-1}$ had excellent long-term survival, while PAH patients with RV fractional area change <36.5% and a peak O_2 pulse <8.0 $mL \cdot beat^{-1}$ had significantly worse survival.

8.3.5 Cardiopulmonary Exercise Testing after Interventions

Very few randomised controlled trials of PAH therapy have included CPET variables as efficacy endpoints [89, 90].

By reducing RV afterload and improving cardiac output and oxygen delivery, PAH therapies such as calcium channel blockers, sildenafil and epoprostenol improve peak $V'O_2$ and ventilatory efficiency [91–93]. Patients who improve peak $V'O_2$, maximal heart rate and O_2 pulse after treatment have better survival, likely due to improvements in cardiac output and stroke volume [86, 94].

In CTEPH, pulmonary endarterectomy is the treatment of choice and involves the surgical removal of obstructing thromboembolic material from the pulmonary arteries. Endarterectomy leads to marked improvements in RV afterload, cardiac function and regained ability to increase stroke volume during exercise, which translates to better exercise capacity and better survival [95–100]. There is also improvement in $V'_E/V'CO_2$ soon after endarterectomy, as a likely result of immediate improvement in cardiac output and a decrease in chemosensitivity, while peak $V'O_2$ continues to improve months after surgery, likely due to rehabilitation and improved peripheral muscle conditioning [101].

Medical therapies approved for PAH are used in inoperable CTEPH patients, which improve exercise capacity and may improve gas exchange and $V'_E/V'CO_2$

[102]. Balloon pulmonary angioplasty (BPA) is another treatment option for inoperable CTEPH patients, which involves dilation of distal obstructing lesions, improving perfusion and lowering mPAP [103, 104]. Right ventricular function and stroke volume improve after BPA, leading to better exercise variables in terms of peak $V'O_2$, $\Delta V'O_2/\Delta WR$, O_2 pulse and $V'_E/V'CO_2$ [105–108]. Oxygen is also a useful intervention to improve exercise performance in patients with pulmonary vascular disease who desaturate during exercise. Supplemental oxygen during exercise increases maximal WR and endurance time and reduces $V'_E/V'CO_2$ by limiting inappropriate chemoreflex-mediated stimulation of V'_E [109].

8.4 Conclusion

Diseases of lung vasculature result from various pathological processes that converge on reducing exercise capacity and lead to early mortality. Understanding the pathophysiological substrates of these outcomes is of upmost importance in order to better orient therapeutic research. But because of the wide range of pathology involved, a one size fits all approach is suboptimal.

Exercise intolerance and dyspnoea in patients with pulmonary vascular disease are multifaceted; the key CPET-related profile responses are a reduced peak $V'O_2$ with impairment of cardiovascular function translating into a reduction in $V'O_2$/WR, low O_2 pulse, and AT, and impaired ventilatory efficiency with altered gas exchange and chemosensitivity. The presence of high $V'_E/V'CO_2$ with a low $P_{ET}CO_2$ in a patient with unexplained dyspnoea should prompt consideration of pulmonary vascular disease in the differential diagnosis and further diagnostic investigations. Abnormal respiratory mechanics and locomotor muscle dysfunction also contribute to dyspnoea, leg fatigue and exercise pathophysiology in many patients. CPET is a useful tool in assessing the degree of functional impairment and disease severity, predicting prognosis and evaluating interventional efficacy.

Key Points

- Cardiopulmonary exercise testing (CPET) in patients with pulmonary vascular diseases may reveal common and non-specific symptoms like dyspnoea and exercise intolerance.
- Dynamic exercise during CPET may provide a greater stress to the right ventricle and pulmonary circulation than static resistive exercises and could thus be more sensitive in detecting an abnormal response in a patient with early pulmonary vascular disease.
- CPET can also help evaluate the severity of disease, gauge responses to treatment and estimate prognosis in patients with known pulmonary vascular disease.
- Even in the absence of significant resting airflow obstruction, dynamic hyperinflation can occur in pulmonary vascular diseases, which contributes to exertional dyspnoea and exercise intolerance.

References

1. Galiè N, Humbert M, Vachiery J-L, et al. ESC/ERS guidelines for the diagnosis and treatment of pulmonary hypertension: the joint task force for the diagnosis and treatment of pulmonary hypertension of the European Society of Cardiology (ESC) and the European Respiratory Society (ERS): endorsed by: Association for European Paediatric and Congenital Cardiology (AEPC), International Society for Heart and Lung Transplantation (ISHLT). Eur Respir J. 2015;46:903–75.
2. Guignabert C, Dorfmüller P. Pathology and pathobiology of pulmonary hypertension. Semin Respir Crit Care Med. 2017;38:571–84.
3. Gall H, Hoeper MM, Richter MJ, Cacheris W, Hinzmann B, Mayer E. An epidemiological analysis of the burden of chronic thromboembolic pulmonary hypertension in the USA, Europe and Japan. Eur Respir Rev. 2017;26:160121.
4. Simonneau G, Torbicki A, Dorfmüller P, Kim N. The pathophysiology of chronic thromboembolic pulmonary hypertension. Eur Respir Rev. 2017;26:160112.
5. Motoji Y, Forton K, Pezzuto B, Faoro V, Naeije R. Resistive or dynamic exercise stress testing of the pulmonary circulation and the right heart. Eur Respir J. 2017;50:1700151.
6. Weatherald J, Farina S, Bruno N, Laveneziana P. Cardiopulmonary exercise testing in pulmonary hypertension. Ann Am Thorac Soc. 2017;14:S84–92.
7. Kovacs G, Olschewski A, Berghold A, Olschewski H. Pulmonary vascular resistances during exercise in normal subjects: a systematic review. Eur Respir J. 2012;39:319–28.
8. Lau EMT, Chemla D, Godinas L, et al. Loss of vascular Distensibility during exercise is an early hemodynamic marker of pulmonary vascular disease. Chest. 2016;149:353–61.
9. Lau EMT, Godinas L, Sitbon O, et al. Resting pulmonary artery pressure of 21-24 mmHg predicts abnormal exercise haemodynamics. Eur Respir J. 2016;47:1436–44.
10. Kovacs G, Avian A, Wutte N, et al. Changes in pulmonary exercise haemodynamics in scleroderma: a 4-year prospective study. Eur Respir J. 2017;50:1601708.
11. Janicki JS, Weber KT, Likoff MJ, Fishman AP. The pressure-flow response of the pulmonary circulation in patients with heart failure and pulmonary vascular disease. Circulation. 1985;72:1270–8.
12. Belenkie I, Dani R, Smith ER, Tyberg JV. Ventricular interaction during experimental acute pulmonary embolism. Circulation. 1988;78:761–8.
13. Belenkie I, Dani R, Smith ER, Tyberg JV. Effects of volume loading during experimental acute pulmonary embolism. Circulation. 1989;80:178–88.
14. Nootens M, Wolfkiel CJ, Chomka EV, Rich S. Understanding right and left ventricular systolic function and interactions at rest and with exercise in primary pulmonary hypertension. Am J Cardiol. 1995;75:374–7.
15. Holverda S, Gan CT-J, Marcus JT, Postmus PE, Boonstra A, Vonk-Noordegraaf A. Impaired stroke volume response to exercise in pulmonary arterial hypertension. J Am Coll Cardiol. 2006;47:1732–3.
16. Naeije R, Badagliacca R. The overloaded right heart and ventricular interdependence. Cardiovasc Res. 2017;113:1474–85.
17. Chemla D, Castelain V, Hoette S, et al. Strong linear relationship between heart rate and mean pulmonary artery pressure in exercising patients with severe precapillary pulmonary hypertension. Am J Physiol Heart Circ Physiol. 2013;305:H769–77.
18. Groepenhoff H, Westerhof N, Jacobs W, Boonstra A, Postmus PE, Vonk-Noordegraaf A. Exercise stroke volume and heart rate response differ in right and left heart failure. Eur J Heart Fail. 2010;12:716–20.
19. Blumberg FC, Arzt M, Lange T, Schroll S, Pfeifer M, Wensel R. Impact of right ventricular reserve on exercise capacity and survival in patients with pulmonary hypertension. Eur J Heart Fail. 2013;15:771–5.
20. van Kan C, van der Plas MN, Reesink HJ, et al. Hemodynamic and ventilatory responses during exercise in chronic thromboembolic disease. J Thorac Cardiovasc Surg. 2016;152:763–71.

21. Sun XG, Hansen JE, Oudiz RJ, Wasserman K. Exercise pathophysiology in patients with primary pulmonary hypertension. Circulation. 2001;104:429–35.
22. Riley MS, Pórszász J, Engelen MP, Brundage BH, Wasserman K. Gas exchange responses to continuous incremental cycle ergometry exercise in primary pulmonary hypertension in humans. Eur J Appl Physiol. 2000;83:63–70.
23. Deboeck G, Niset G, Lamotte M, Vachiéry JL, Naeije R. Exercise testing in pulmonary arterial hypertension and in chronic heart failure. Eur Respir J. 2004;23:747–51.
24. Potus F, Malenfant S, Graydon C, et al. Impaired angiogenesis and peripheral muscle microcirculation loss contribute to exercise intolerance in pulmonary arterial hypertension. Am J Respir Crit Care Med. 2014;190:318–28.
25. Mainguy V, Maltais F, Saey D, et al. Peripheral muscle dysfunction in idiopathic pulmonary arterial hypertension. Thorax. 2010;65:113–7.
26. Wensel R, Opitz CF, Anker SD, et al. Assessment of survival in patients with primary pulmonary hypertension: importance of cardiopulmonary exercise testing. Circulation. 2002;106:319–24.
27. Imai K, Sato H, Hori M, et al. Vagally mediated heart rate recovery after exercise is accelerated in athletes but blunted in patients with chronic heart failure. J Am Coll Cardiol. 1994;24:1529–35.
28. Ramos RP, Arakaki JSO, Barbosa P, et al. Heart rate recovery in pulmonary arterial hypertension: relationship with exercise capacity and prognosis. Am Heart J. 2012;163:580–8.
29. Riley MS, Pórszász J, Engelen MP, Shapiro SM, Brundage BH, Wasserman K. Responses to constant work rate bicycle ergometry exercise in primary pulmonary hypertension: the effect of inhaled nitric oxide. J Am Coll Cardiol. 2000;36:547–56.
30. Theodore J, Robin ED, Morris AJ, et al. Augmented ventilatory response to exercise in pulmonary hypertension. Chest. 1986;89:39–44.
31. D'Alonzo GE, Gianotti LA, Pohil RL, et al. Comparison of progressive exercise performance of normal subjects and patients with primary pulmonary hypertension. Chest. 1987;92:57–62.
32. Sun X-G, Hansen JE, Oudiz RJ, Wasserman K. Pulmonary function in primary pulmonary hypertension. J Am Coll Cardiol. 2003;41:1028–35.
33. Laveneziana P, Garcia G, Joureau B, et al. Dynamic respiratory mechanics and exertional dyspnoea in pulmonary arterial hypertension. Eur Respir J. 2013;41:578–87.
34. Meyer FJ, Ewert R, Hoeper MM, et al. Peripheral airway obstruction in primary pulmonary hypertension. Thorax. 2002;57:473–6.
35. Richter MJ, Voswinckel R, Tiede H, et al. Dynamic hyperinflation during exercise in patients with precapillary pulmonary hypertension. Respir Med. 2012;106:308–13.
36. de Man FS, van Hees HWH, Handoko ML, et al. Diaphragm muscle fiber weakness in pulmonary hypertension. Am J Respir Crit Care Med. 2011;183:1411–8.
37. Manders E, Bonta PI, Kloek JJ, et al. Reduced force of diaphragm muscle fibers in patients with chronic thromboembolic pulmonary hypertension. Am J Physiol Lung Cell Mol Physiol. 2016;311:L20–8.
38. Meyer FJ, Lossnitzer D, Kristen AV, et al. Respiratory muscle dysfunction in idiopathic pulmonary arterial hypertension. Eur Respir J. 2005;25:125–30.
39. Laveneziana P, Humbert M, Godinas L, et al. Inspiratory muscle function, dynamic hyperinflation and exertional dyspnoea in pulmonary arterial hypertension. Eur Respir J. 2015;45:1495–8.
40. Sun X-G, Hansen JE, Garatachea N, Storer TW, Wasserman K. Ventilatory efficiency during exercise in healthy subjects. Am J Respir Crit Care Med. 2002;166:1443–8.
41. Vicenzi M, Deboeck G, Faoro V, Loison J, Vachiery J-L, Naeije R. Exercise oscillatory ventilation in heart failure and in pulmonary arterial hypertension. Int J Cardiol. 2016;202:736–40.
42. Reybrouck T, Mertens L, Schulze-Neick I, et al. Ventilatory inefficiency for carbon dioxide during exercise in patients with pulmonary hypertension. Clin Physiol. 1998;18:337–44.
43. Liu W-H, Luo Q, Liu Z-H, et al. Pulmonary function differences in patients with chronic right heart failure secondary to pulmonary arterial hypertension and chronic left heart failure. Med Sci Monit. 2014;20:960–6.
44. McCabe C, Deboeck G, Harvey I, et al. Inefficient exercise gas exchange identifies pulmonary hypertension in chronic thromboembolic obstruction following pulmonary embolism. Thromb Res. 2013;132:659–65.

45. Xi Q, Zhao Z, Liu Z, Ma X, Luo Q, Liu W. The lowest VE/VCO_2 ratio best identifies chronic thromboembolic pulmonary hypertension. Thromb Res. 2014;134:1208–13.
46. Zhai Z, Murphy K, Tighe H, et al. Differences in ventilatory inefficiency between pulmonary arterial hypertension and chronic thromboembolic pulmonary hypertension. Chest. 2011;140:1284–91.
47. Godinas L, Sattler C, Lau EM, et al. Dead-space ventilation is linked to exercise capacity and survival in distal chronic thromboembolic pulmonary hypertension. J Heart Lung Transplant. 2017;36:1234–42.
48. Yasunobu Y, Oudiz RJ, Sun X-G, Hansen JE, Wasserman K. End-tidal PCO2 abnormality and exercise limitation in patients with primary pulmonary hypertension. Chest. 2005;127:1637–46.
49. Markowitz DH, Systrom DM. Diagnosis of pulmonary vascular limit to exercise by cardiopulmonary exercise testing. J Heart Lung Transplant. 2004;23:88–95.
50. Weatherald J, Sattler C, Garcia G, Laveneziana P. Ventilatory response to exercise in cardiopulmonary disease: the role of Chemosensitivity and dead space. Eur Respir J. 2018;51:1700860.
51. Velez-Roa S, Ciarka A, Najem B, Vachiery J-L, Naeije R, van de Borne P. Increased sympathetic nerve activity in pulmonary artery hypertension. Circulation. 2004;110:1308–12.
52. Naeije R, van de Borne P. Clinical relevance of autonomic nervous system disturbances in pulmonary arterial hypertension. Eur Respir J. 2009;34:792–4.
53. Wensel R, Jilek C, Dörr M, et al. Impaired cardiac autonomic control relates to disease severity in pulmonary hypertension. Eur Respir J. 2009;34:895–901.
54. Hoeper MM, Pletz MW, Golpon H, Welte T. Prognostic value of blood gas analyses in patients with idiopathic pulmonary arterial hypertension. Eur Respir J. 2007;29:944–50.
55. Weatherald J, Sattler C, Boucly A, et al. Chemosensitivity and ventilatory inefficiency in pulmonary vascular diseases. Eur Respir J. 2017;50:1966.
56. Jones PW, Huszczuk A, Wasserman K. Cardiac output as a controller of ventilation through changes in right ventricular load. J Appl Physiol. 1982;53:218–24.
57. Ciarka A, Doan V, Velez-Roa S, Naeije R, van de Borne P. Prognostic significance of sympathetic nervous system activation in pulmonary arterial hypertension. Am J Respir Crit Care Med. 2010;181:1269–75.
58. Dimopoulos S, Anastasiou-Nana M, Katsaros F, et al. Impairment of autonomic nervous system activity in patients with pulmonary arterial hypertension: a case control study. J Card Fail. 2009;15:882–9.
59. Ponikowski PP, Chua TP, Francis DP, Capucci A, Coats AJ, Piepoli MF. Muscle ergoreceptor overactivity reflects deterioration in clinical status and cardiorespiratory reflex control in chronic heart failure. Circulation. 2001;104:2324–30.
60. Sun X-G, Hansen JE, Oudiz RJ, Wasserman K. Gas exchange detection of exercise-induced right-to-left shunt in patients with primary pulmonary hypertension. Circulation. 2002;105:54–60.
61. Chua TP, Clark AL, Amadi AA, Coats AJ. Relation between chemosensitivity and the ventilatory response to exercise in chronic heart failure. J Am Coll Cardiol. 1996;27:650–7.
62. Scott AC, Wensel R, Davos CH, et al. Skeletal muscle reflex in heart failure patients: role of hydrogen. Circulation. 2003;107:300–6.
63. Ciarka A, Vachièry J-L, Houssière A, et al. Atrial septostomy decreases sympathetic overactivity in pulmonary arterial hypertension. Chest. 2007;131:1831–7.
64. Deboeck G, Scoditti C, Huez S, et al. Exercise testing to predict outcome in idiopathic versus associated pulmonary arterial hypertension. Eur Respir J. 2012;40:1410–9.
65. Schwaiblmair M, Faul C, von Scheidt W, Berghaus TM. Ventilatory efficiency testing as prognostic value in patients with pulmonary hypertension. BMC Pulm Med. 2012;12:23.
66. Scheidl SJ, Englisch C, Kovacs G, et al. Diagnosis of CTEPH versus IPAH using capillary to end-tidal carbon dioxide gradients. Eur Respir J. 2012;39:119–24.
67. Held M, Grün M, Holl R, et al. Cardiopulmonary exercise testing to detect chronic thromboembolic pulmonary hypertension in patients with normal echocardiography. Respiration. 2014;87:379–87.
68. Laveneziana P, Montani D, Dorfmüller P, et al. Mechanisms of exertional dyspnoea in pulmonary veno-occlusive disease with EIF2AK4 mutations. Eur Respir J. 2014;44:1069–72.

69. Wasserman K, Zhang YY, Gitt A, et al. Lung function and exercise gas exchange in chronic heart failure. Circulation. 1997;96:2221–7.
70. Houstis NE, Eisman AS, Pappagianopoulos PP, et al. Exercise intolerance in HFpEF: diagnosing and ranking its causes using personalized O2 pathway analysis. Circulation. 2018;137:148–61.
71. Van Iterson EH, Johnson BD, Borlaug BA, Olson TP. Physiological dead space and arterial carbon dioxide contributions to exercise ventilatory inefficiency in patients with reduced or preserved ejection fraction heart failure. Eur J Heart Fail. 2017;19:1675–85.
72. Dantzker DR, D'Alonzo GE, Bower JS, Popat K, Crevey BJ. Pulmonary gas exchange during exercise in patients with chronic obliterative pulmonary hypertension. Am Rev Respir Dis. 1984;130:412–6.
73. Wagner PD. The physiological basis of pulmonary gas exchange: implications for clinical interpretation of arterial blood gases. Eur Respir J. 2015;45:227–43.
74. Oudiz RJ, Midde R, Hovenesyan A, et al. Usefulness of right-to-left shunting and poor exercise gas exchange for predicting prognosis in patients with pulmonary arterial hypertension. Am J Cardiol. 2010;105:1186–91.
75. Jones NL, Robertson DG, Kane JW. Difference between end-tidal and arterial PCO2 in exercise. J Appl Physiol. 1979;47:954–60.
76. Liu Z, Vargas F, Stansbury D, Sasse SA, Light RW. Comparison of the end-tidal arterial PCO2 gradient during exercise in normal subjects and in patients with severe COPD. Chest. 1995;107:1218–24.
77. Esposito F, Mathieu-Costello O, Shabetai R, Wagner PD, Richardson RS. Limited maximal exercise capacity in patients with chronic heart failure: partitioning the contributors. J Am Coll Cardiol. 2010;55:1945–54.
78. Dimopoulos S, Tzanis G, Manetos C, et al. Peripheral muscle microcirculatory alterations in patients with pulmonary arterial hypertension: a pilot study. Respir Care. 2013;58:2134–41.
79. Tolle J, Waxman A, Systrom D. Impaired systemic oxygen extraction at maximum exercise in pulmonary hypertension. Med Sci Sports Exerc. 2008;40:3–8.
80. Bauer R, Dehnert C, Schoene P, et al. Skeletal muscle dysfunction in patients with idiopathic pulmonary arterial hypertension. Respir Med. 2007;101:2366–9.
81. Duscha BD, Robbins JL, Jones WS, et al. Angiogenesis in skeletal muscle precede improvements in peak oxygen uptake in peripheral artery disease patients. Arterioscler Thromb Vasc Biol. 2011;31:2742–8.
82. Esposito F, Reese V, Shabetai R, Wagner PD, Richardson RS. Isolated quadriceps training increases maximal exercise capacity in chronic heart failure: the role of skeletal muscle convective and diffusive oxygen transport. J Am Coll Cardiol. 2011;58:1353–62.
83. Mereles D, Ehlken N, Kreuscher S, et al. Exercise and respiratory training improve exercise capacity and quality of life in patients with severe chronic pulmonary hypertension. Circulation. 2006;114:1482–9.
84. Ehlken N, Lichtblau M, Klose H, et al. Exercise training improves peak oxygen consumption and haemodynamics in patients with severe pulmonary arterial hypertension and inoperable chronic thrombo-embolic pulmonary hypertension: a prospective, randomized, controlled trial. Eur Heart J. 2016;37:35–44.
85. Puente-Maestu L, Palange P, Casaburi R, et al. Use of exercise testing in the evaluation of interventional efficacy: an official ERS statement. Eur Respir J. 2016;47:429–60.
86. Groepenhoff H, Vonk-Noordegraaf A, van de Veerdonk MC, Boonstra A, Westerhof N, Bogaard HJ. Prognostic relevance of changes in exercise test variables in pulmonary arterial hypertension. PLoS One. 2013;8:e72013.
87. Wensel R, Francis DP, Meyer FJ, et al. Incremental prognostic value of cardiopulmonary exercise testing and resting haemodynamics in pulmonary arterial hypertension. Int J Cardiol. 2013;167:1193–8.
88. Badagliacca R, Papa S, Valli G, et al. Echocardiography combined with cardiopulmonary exercise testing for the prediction of outcome in idiopathic pulmonary arterial hypertension. Chest. 2016;150:1313–22.
89. Barst RJ, McGoon M, McLaughlin V, et al. Beraprost therapy for pulmonary arterial hypertension. J Am Coll Cardiol. 2003;41:2119–25.

90. Barst RJ, Langleben D, Frost A, et al. Sitaxsentan therapy for pulmonary arterial hypertension. Am J Respir Crit Care Med. 2004;169:441–7.
91. Tang Y, Yao L, Liu Z, et al. Effect of calcium channel blockers evaluated by cardiopulmonary exercise testing in idiopathic pulmonary arterial hypertension responding to acute pulmonary vasoreactivity testing. Pulm Pharmacol Ther. 2017;43:26–31.
92. Oudiz RJ, Roveran G, Hansen JE, Sun X-G, Wasserman K. Effect of sildenafil on ventilatory efficiency and exercise tolerance in pulmonary hypertension. Eur J Heart Fail. 2007;9:917–21.
93. Wax D, Garofano R, Barst RJ. Effects of long-term infusion of prostacyclin on exercise performance in patients with primary pulmonary hypertension. Chest. 1999;116:914–20.
94. Weatherald J, Boucly A, Chemla D, et al. The prognostic value of follow-up hemodynamic variables after initial management in pulmonary arterial hypertension. Circulation. 2018;137:693–704.
95. Surie S, van der Plas MN, Marcus JT, et al. Effect of pulmonary endarterectomy for chronic thromboembolic pulmonary hypertension on stroke volume response to exercise. Am J Cardiol. 2014;114:136–40.
96. Taboada D, Pepke-Zaba J, Jenkins DP, et al. Outcome of pulmonary endarterectomy in symptomatic chronic thromboembolic disease. Eur Respir J. 2014;44:1635–45.
97. Claessen G, La Gerche A, Dymarkowski S, Claus P, Delcroix M, Heidbuchel H. Pulmonary vascular and right ventricular reserve in patients with normalized resting hemodynamics after pulmonary endarterectomy. J Am Heart Assoc. 2015;4:e001602.
98. Richter MJ, Sommer N, Gall H, et al. Pulmonary hemodynamic response to exercise in chronic thromboembolic pulmonary hypertension before and after pulmonary endarterectomy. Respiration. 2015;90:63–73.
99. Bonderman D, Martischnig AM, Vonbank K, et al. Right ventricular load at exercise is a cause of persistent exercise limitation in patients with normal resting pulmonary vascular resistance after pulmonary endarterectomy. Chest. 2011;139:122–7.
100. Delcroix M, Lang I, Pepke-Zaba J, et al. Long-term outcome of patients with chronic thromboembolic pulmonary hypertension: results from an international prospective registry. Circulation. 2016;133:859–71.
101. Iwase T, Nagaya N, Ando M, et al. Acute and chronic effects of surgical thromboendarterectomy on exercise capacity and ventilatory efficiency in patients with chronic thromboembolic pulmonary hypertension. Heart. 2001;86:188–92.
102. Charalampopoulos A, Gibbs JSR, Davies RJ, et al. Exercise physiological responses to drug treatments in chronic thromboembolic pulmonary hypertension. J Appl Physiol (1985). 2016;121:623–8.
103. Lang I, Meyer BC, Ogo T, et al. Balloon pulmonary angioplasty in chronic thromboembolic pulmonary hypertension. Eur Respir Rev. 2017;26:160119.
104. Inami T, Kataoka M, Yanagisawa R, et al. Long-term outcomes after percutaneous transluminal pulmonary angioplasty for chronic thromboembolic pulmonary hypertension. Circulation. 2016;134:2030–2.
105. Fukui S, Ogo T, Goto Y, et al. Exercise intolerance and ventilatory inefficiency improve early after balloon pulmonary angioplasty in patients with inoperable chronic thromboembolic pulmonary hypertension. Int J Cardiol. 2015;180:66–8.
106. Fukui S, Ogo T, Morita Y, et al. Right ventricular reverse remodelling after balloon pulmonary angioplasty. Eur Respir J. 2014;43:1394–402.
107. Andreassen AK, Ragnarsson A, Gude E, Geiran O, Andersen R. Balloon pulmonary angioplasty in patients with inoperable chronic thromboembolic pulmonary hypertension. Heart. 2013;99:1415–20.
108. Akizuki M, Serizawa N, Ueno A, Adachi T, Hagiwara N. Effect of balloon pulmonary angioplasty on respiratory function in patients with chronic thromboembolic pulmonary hypertension. Chest. 2017;151:643–9.
109. Ulrich S, Hasler ED, Saxer S, et al. Effect of breathing oxygen-enriched air on exercise performance in patients with precapillary pulmonary hypertension: randomized, sham-controlled cross-over trial. Eur Heart J. 2017;38:1159–68.

Respiratory Infections

9

Marcin Kurowski

Abstract

Respiratory infections are the most frequent cause of athletes' visits in medical practices. Simultaneously, contradictory beliefs are generally held regarding influence of repeated exercise on immunity reflected in susceptibility to infections. Many tend to claim that exercise weakens the immunity and renders regular exercisers more prone to develop airway infections. On the other hand, voices are heard in favor of the possible beneficial influence of regular exercise on immune system efficiency. This chapter focuses on the associations of various kinds of exercise with respiratory infection susceptibility. Influence of exercise, in particular associated with competitive performance, on selected innate and acquired immune response mechanisms is also addressed.

9.1 General Overview

Some scientific evidence suggests that acute bouts of exercise are followed with period of increased susceptibility to respiratory infections [1–4]. Approximately, a twofold increase in prevalence of upper respiratory tract infection (URTI) symptoms after long-distance run is observed, as compared with non-runner counterparts from control groups [3, 5]. On the other hand, regular moderate physical activity is believed to contribute to greater resistance to respiratory infections. Regular exercise at moderate level is associated with reduced incidence of URTI episodes and lower intensity of infectious symptoms [6, 7]. This phenomenon has been graphically depicted as the so-called J-shaped curve which clearly shows that both sedentary

M. Kurowski (✉)
Department of Immunology and Allergy, Medical University of Łódź, Łódź, Poland
e-mail: marcin.kurowski@umed.lodz.pl

A. Cogo et al. (eds.), *Exercise and Sports Pulmonology*,
https://doi.org/10.1007/978-3-030-05258-4_9

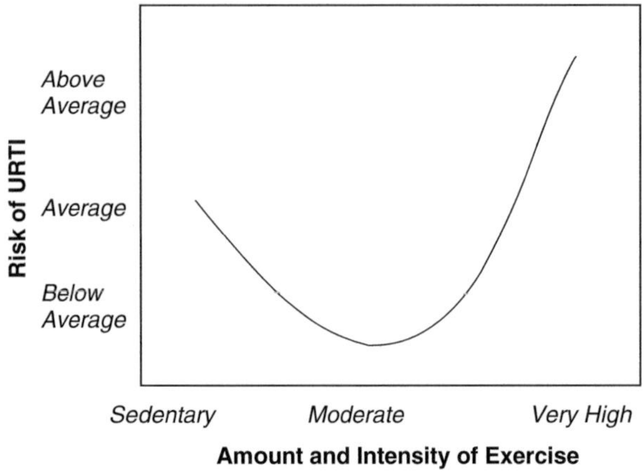

Fig. 9.1 "J"-shaped model of relationship between varying amounts of exercise and risk of URTI. This model suggests that moderate exercise may lower risk of respiratory infection while excessive amounts may increase the risk. Source: Nieman D. Exercise, upper respiratory tract infection, and the immune system. Med Sci Sports Exerc 1994; 26 (2): 128–39. Ref. [1]

lifestyle and excessive strenuous exercise constitute risk factors for increased susceptibility to respiratory infections [1] (Fig. 9.1).

Some reports from studies in professional athletes suggest, however, that regular high-level exercise tends to be protective with regard to URTI symptoms [8, 9]. These data may seem to question the "J-curve" model, but it is hypothesized that presence of other, not yet identified, genetic or behavioral factors might influence one's susceptibility to infection in a high-level exercise context [10]. This modified model has been nicknamed "S-curve" [9]. One possible explanation for lower infection rate in top-level exercisers is the fact that subjects engaging in regular exercise of high intensity are predominantly those with perfect physical stamina and perfect immune health status that enable them to resist infections and carry on with training, thus permitting their inclusion to "elite" group on the basis of their achievements.

URTI usually resolve in a few days. The "check-neck rule" is commonly applied to decide whether the athletes should be stopped from training and competition (no stop for symptoms above the neck, stop in case of cough, fever, etc.). Sore throat, a frequent symptom of URTI, should alert to exclude mononucleosis and potential myocardial complications of β-hemolytic streptococci.

Competitive athletes frequently report URTI symptoms, but their infectious etiology cannot be confirmed in a considerable proportion of cases. Both viruses (rhinovirus, particularly in spring, influenza and parainfluenza virus, adenovirus and coronavirus) and bacteria (streptococci, staphylococci) are claimed to be involved in causing URTI in athletes, although data from large systematic studies including the isolation and identification of the responsible agents are still very limited (Table 9.1). Spence and colleagues have analyzed 37 infectious episodes in 28 competitive and noncompetitive athletes [11]. In that group, infectious etiology could be established

Table 9.1 Pathogens most frequently identified as causative agents for URTI symptoms in athletes

Viral
Rhinovirus
Influenza virus
Parainfluenza virus 1, 2, and 3
Adenovirus
Coronavirus
Bacterial
M. pneumoniae
S. aureus
S. pneumoniae

Basing on Refs. [11, 12]

only with regard to 11 episodes. Identified viral pathogens included rhinoviruses and adenoviruses, whereas *M. pneumoniae*, *S. aureus*, and *S. pneumoniae* were confirmed as bacteria causing URTI symptoms. A study by Cox et al. [12] identified viral or bacterial pathogen in 30% of athletes with upper respiratory symptoms, while 57% of all cases were accompanied either by presence of a detectable pathogen or by other laboratory pathological features of infection. Bacterial cause was found in 3% of cases only, while most frequently detected viral pathogens included rhinovirus (10%); influenza virus (10%); parainfluenza viruses 1, 2, and 3 (6%); and coronaviruses (3%). Hitherto published data indicate that pathogen identification is possible in approx. 45% of URTI symptoms in athletes [13]. A similar proportion has been also described regarding URTI symptoms in the general population [14].

9.2 Immunological Changes Related to Exercise

9.2.1 Immunoglobulins and Exercise

Decreased effectiveness of local humoral immune response at the level of the respiratory mucosae is one of the characteristic features associated with intense physical exercise. This is predominantly reflected in decreased salivary IgA (sIgA) concentrations, but the importance of other proteins contained in saliva has been recently pointed at in the context of susceptibility to upper respiratory tract infections [15]. Increased URTI susceptibility following participation in long-distance runs was observed in multiple studies [1–5]. Already in 1983 Peters and Bateman [3] ascertained the presence of URTI symptoms in one-third of ultramarathon participants as compared with 15% of non-running controls. Moreover, the prevalence of URTI symptoms was more prevalent in those who performed better in terms of time required to complete the run. Another study performed in runners showed, however, that those who were more committed to intensive training during preparations to long-distance run were less likely to develop infection symptoms [4]. Robson-Ansley and colleagues [5] reported a 47% prevalence of URTI symptoms over the 2-week period following participation to the London Marathon. In their study, however, a significant correlation of URTI symptoms occurred with positive IgE-mediated allergen sensitizations and atopic status, as assessed by the AQUA questionnaire designed for allergy screening in athletes [16]. The question related to the proportion of upper respiratory infection-like symptoms due to exposure to inhalant allergens and not to infection per se therefore arises. Several studies were performed in nonprofessional exercisers or sedentary subjects [6, 7] revealing that decrease in susceptibility to URTI is predominantly visible in regular exercisers who do not advance into professional, high-endurance type of training.

Shifts in salivary IgA are observed in subjects loaded with strenuous exercise associated with sports training [17–22], as well as being part of military training curriculum [23–25]. Decreased salivary IgA is usually correlated with increased URTI susceptibility, although this mutual association is not always evident [15]. A pioneer study by Tomasi et al. [18] comparing salivary IgA levels in cross-country skiers versus age-matched controls showed significant decreases both at baseline and after participation in a competitive run. Studies in swimmers [19, 20] also showed decreased IgA levels in saliva after training. Additionally, salivary IgA levels were inversely correlated with duration and intensity of training, as well as with number of reported infections [19]. Similarly, decreased salivary IgA was described as a risk factor for developing upper respiratory symptoms in male yacht race participants [21].

There exist, however, studies that failed to show modification of salivary IgA in response to a repeated training exercise [26, 27] pertinent to different sports disciplines. Moreover, some studies do not confirm direct association between decreased salivary IgA and frequency or severity of respiratory infections [28–30]. This leads to assumption that influence of factors other than excessive exercise load should be taken into consideration while interpreting the data regarding salivary IgA levels. These factors include, among others, pattern of exercise, its intensity and duration, as well as the general stamina of each individual. In the case of extremely intensive training programs, provisions should be made for additional factors modifying immune response, such as increased energy consumption, sleep deprivation, altitude above sea level, and stress-associated psychological factors [15, 31–33].

Moderate physical activity as part of lifestyle modification plans in previously sedentary subjects leads to increase in salivary IgA levels. This fact speaks in favor of beneficial anti-inflammatory and immunomodulatory properties attributed to moderate but regular exercise [34, 35]. In young soccer players, it has been observed that salivary IgA increase considerably during a 2-week detraining period scheduled at the end of a 21-week competitive season. Increase in salivary IgA was accompanied by a significant decrease in URTI symptom score [36].

Data regarding serum immunoglobulins are contradictory. According to many authors, IgG levels in endurance athletes increase both directly after exercise and over longer observation periods [37–40]. In other studies, however, considerable falls in serum IgG were observed in association with prolonged strenuous exercise (75-kilometer run, 3-week rugby training camp participation, 2 weeks of regular running training) [41–44].

Similarly ambiguous results have been observed regarding serum IgM where both reductions [37, 41–43] and increases [38, 45] were seen under intensive exercise conditions.

9.3 Cytokines, Inflammation, and URTI Susceptibility in Exercisers and Athletes

Physical exercise is a stimulus inducing multiple cytokine synthesis and release. At the same time, as mentioned above, it may have anti-inflammatory and immunomodulatory properties. In spite of an anti-inflammatory action, however, an acute bout of exercise results in increased release of acute phase proteins and inflammation cytokines (IL-6, IL-1β, MIP-1α, IL-8). Secondary to this, increased serum levels of anti-inflammatory cytokines have been observed (e.g., IL-10, IL-1ra) [15, 46–51]. Serum levels of anti-inflammatory IL-1ra protein observed in young speed skaters were negatively correlated with ambient air temperature during winter season in the outdoor training area, suggesting that serum IL-1ra is partly reflecting cold air exposure in winter athletes. In addition, serum IL-1ra was significantly elevated in winter season only in athletes not reporting frequent respiratory tract infections [52] which adds further insight into the significance of anti-inflammatory cytokine properties in conditioning susceptibility to respiratory infections.

Intensive exercise is accompanied by hyperventilation resulting in sequential warming and cooling of the airways. This may lead to epithelial cells dehydration and, subsequently, hyperosmolar stress causing increased release of chemokines, such as RANTES and IL-8 which, in turn, induce leukocyte influx to the airways.

As a result of decrease in Th1 cell numbers with Th2 numbers remaining at constant level, the Th1/Th2 balance is skewed toward Th2 predominance due to strenuous exercise. Such Th1/Th2 unbalance plays an important role in increasing URTI susceptibility. Cortisol and adrenaline released during exercise decrease the intensity of Th1-dependent response. Muscle fiber-derived IL-6 directly stimulates Th2-mediated response which—together with decreased Th1 response—is likely to contribute to lower efficacy of antiviral response mechanisms [53, 54].

It is hypothesized that persistent low-grade inflammation that is currently considered important in pathogenesis of, among others, cardiovascular and metabolic diseases may be present in the airways of subjects performing regular intensive exercise. Indeed, there is an increased flood of inflammatory cells (neutrophils, eosinophils, lymphocytes) in the athletes' airways resulting from a hyperventilation-induced increase in airway osmolarity stimulating bronchial epithelial cells to release chemotactic factors. It has been, however, also observed that exercise bout leads to considerable lowering of the expression of adhesive particles on the surface of inflammatory cells explaining why airway inflammation may appear blunted in athletes in spite of the presence of multiple exercise-associated pro-inflammatory stimuli. The appearance of infection-like episodes without evidence for the role of infectious pathogens may result from a transient loss of control of such local inflammation due to external physicochemical factors present in the environment where exercise is performed (e.g., swimming pool disinfectants, cold and dry air hyperpnea) [13].

Intensity of inflammation is also associated with anti-inflammatory and immunomodulatory function of the Clara cell proteins (CC16) which has immunomodulatory and anti-inflammatory properties. Assessment of serum CC16 levels can also be a marker of airway epithelium damage. In Olympic athletes (as compared with a general population sample) significantly lower serum CC16 levels were ascertained, irrespective of the discipline performed or the training regime followed [55]. In addition, athletes reporting frequent respiratory infections had significantly lower serum CC16 as compared to illness-resistant athletes, and a serum CC16 level below 5 ng/mL was associated with more than twofold increase of frequent URTI risk.

9.4 Conclusions

Regular exercise of moderate intensity remains the best lifestyle intervention if long-term prophylaxis of respiratory infections and lowering of inflammation are aimed at. However, repeated and strenuous exercise may predispose to the development of respiratory tract infections. Careful screening of athletes, also including the use of questionnaires specifically designed, appears highly useful in managing those in whom respiratory symptoms impair sports performance.

Various aspects of changes in lifestyle and everyday routine, such as diet, sleep deprivation, traveling, jet lag, stress, community living, and general health status, should be considered as modifiers of URTI susceptibility even in apparently illness-prone subjects. Moreover, concomitant asthma, allergic diseases, and atopy may modify the susceptibility to respiratory infections.

In the case of respiratory symptoms in athletes, differential diagnoses should be also always considered (i.e., exercise-induced bronchoconstriction, vocal-cord dysfunction).

Several dietary supplements have been reported to restore a normal immune function and reduce the increased risk of URTI in athletes, including Vitamin C, glucose, lipids, zinc, *Echinacea*, and other minerals. However, data available are scarce and sometimes conflicting; systematic reviews and meta-analysis failed to show any effect for most of the remedies suggested [56–61].

Lastly, the high prevalence of URTI in athletes has induced some authors to recommend yearly vaccinations against influenza (and hepatitis A and B immunization) [62].

Key Points

- Regular moderate physical activity is believed to contribute to a more effective immune response against respiratory infection agents.
- Repeated and strenuous exercise predisposes to the development of symptoms suggestive of respiratory tract infections.

- Intensive exercise training leads to decrease in salivary IgA which is frequently associated with an increased prevalence of respiratory symptoms.
- Pathogens responsible for athletes' upper respiratory tract infections (URTI) can be identified in less than half of the episodes.

References

1. Nieman D. Exercise, upper respiratory tract infection, and the immune system. Med Sci Sports Exerc. 1994;26(2):128–39.
2. Nieman D, Johanssen L, Lee J, Arabatzis K. Infectious episodes in runners before and after the Los Angeles Marathon. J Sports Med Phys Fitness. 1990;30(3):316–28.
3. Peters E, Bateman E. Ultramarathon running and upper respiratory tract infections. An epidemiological survey. S Afr Med J. 1983;64(15):582–4.
4. Nieman D, Johanssen L, Lee J. Infectious episodes in runners before and after a roadrace. J Sports Med Phys Fitness. 1989;29(3):289–96.
5. Robson-Ansley P, Howatson G, Tallent J, Mitcheson K, Walshe I, Toms C, et al. Prevalence of allergy and upper respiratory tract symptoms in runners of London Marathon. Med Sci Sports Exerc. 2012;44(6):999–1004.
6. Kostka T, Drygas W, Jegier A, Praczko K. Physical activity and upper respiratory tract infections. Int J Sports Med. 2008;29(2):158–62.
7. Nieman DC, Henson DA, Austin MD, Sha W. Upper respiratory tract infection is reduced in physically fit and active adults. Br J Sports Med. 2011;45(12):987–92.
8. Fondell E, Lagerros YT, Sundberg CJ, Lekander M, Bälter O, Rothman KJ, et al. Physical activity, stress, and self-reported upper respiratory tract infection. Med Sci Sports Exerc. 2011;43(2):272–9.
9. Malm C. Susceptibility to infections in elite athletes: the S-curve. Scand J Med Sci Sports. 2006;16(1):4–6.
10. Walsh N, Oliver S. Exercise, immune function and respiratory infection: an update on the influence of training and environmental stress. Immunol Cell Biol. 2016;94(2):132–9.
11. Spence L, Brown W, Pyne D, Nissen M, Sloots T, McCormack J, et al. Incidence, etiology, and symptomatology of upper respiratory illness in elite athletes. Med Sci Sports Exerc. 2007;39(4):577–86.
12. Cox AJ, Gleeson M, Pyne DB, Callister R, Hopkins WG, Fricker PA. Clinical and laboratory evaluation of upper respiratory symptoms in elite athletes. Clin J Sport Med. 2008;18(5): 438–45.
13. Bermon S. Airway inflammation and upper respiratory tract infection in athletes: is there a link? Exerc Immunol Rev. 2007;13:6–14.
14. Mäkelä MJ, Puhakka T, Ruuskanen O, Leinonen M, Saikku P, Kimpimäki M, et al. Viruses and Bacteria in the etiology of the common cold. J Clin Microbiol. 1998;36(2):539–42.
15. Walsh N, Gleeson M, Shephard R, Gleeson M, Woods J, Bishop N, et al. Position statement. Part one: immune function and exercise. Exerc Immunol Rev. 2011;17:6–63.
16. Bonini M, Braido F, Baiardini I, Del Giacco S, Gramiccioni C, Manara M, et al. AQUA: allergy questionnaire for athletes. Development and validation. Med Sci Sports Exerc. 2009;41(5):1034–41.
17. Fahlman M, Engels H. Mucosal IgA and URTI in American college football players: a year longitudinal study. Med Sci Sports Exerc. 2005;37:374–80.
18. Tomasi T, Trudeau F, Czerwinski D, Erredge S. Immune parameters in athletes before and after strenuous exercise. J Clin Immunol. 1982;2:173–8.

19. Gleeson M, McDonald W, Pyne D, Cripps A, Francis J, Fricker P, et al. Salivary IgA levels and infection risk in elite swimmers. Med Sci Sports Exerc. 1999;31(1):67–73.
20. Gleeson M, McDonald WA, Cripps AW, Pyne DB, Clancy RL, Fricker PA. The effect on immunity of long-term intensive training in elite swimmers. Clin Exp Immunol. 1995;102(1):210–6.
21. Neville V, Gleeson M, Folland JP. Salivary IgA as a risk factor for upper respiratory infections in elite professional athletes. Med Sci Sports Exerc. 2008;40(7):1228–36.
22. Mortatti AL, Moreira A, Aoki MS, Crewther BT, Castagna C, de Arruda AFS, et al. Effect of competition on salivary cortisol, immunoglobulin A, and upper respiratory tract infections in elite young soccer players. J Strength Cond Res. 2012;26(5):1396–401.
23. Carins J, Booth C. Salivary immunoglobulin-A as a marker of stress during strenuous physical training. Aviat Space Environ Med. 2002;73:1203–7.
24. Tiollier E, Gomez-Merino D, Burnat P, Jouanin J, Bourrilhon C, Filaire E, et al. Intense training: mucosal immunity and incidence of respiratory infections. Eur J Appl Physiol. 2005;93:421–8.
25. Whitham M, Laing S, Dorrington M, Walters R, Dunklin S, Bland D, et al. The influence of an arduous military training program on immune function and upper respiratory tract infection incidence. Mil Med. 2006;171:703–9.
26. Vardiman JP, Riggs CE, Galloway DL, Waxman MB, Touchberry CD, Gallagher PM. Salivary IgA is not a reliable Indicator of upper respiratory infection in collegiate female soccer athletes. J Strength Cond Res. 2011;25(7):1937–42.
27. Moreira A, Bacurau RFP, Napimoga MH, Arruda AFS, Freitas CG, Drago G, et al. Salivary IL-21 and IGA responses to a competitive match in elite basketball players. Biol Sport. 2013;30(4):243–7.
28. Alexandre M, Franco A, Ramos CP, Clóvis F, Carlos SA, Roberto DOP, et al. The impact of a 17-day training period for an international championship on mucosal immune parameters in top-level basketball players and staff members. Eur J Oral Sci. 2008;116(5):431–7.
29. Milanez VF, Ramos SP, Okuno NM, Boullosa DA, Nakamura FY. Evidence of a non-linear dose-response relationship between training load and stress markers in elite female futsal players. J Sports Sci Med. 2014;13(1):22–9.
30. Moraes H, Aoki MS, Freitas CG, Arruda AFS, Drago G, Moreira A. SIgA response and incidence of upper respiratory tract infections during intensified training in youth basketball players. Biol Sport. 2017;34(1):49–55.
31. Oliver S, Laing S, Wilson S, Bilzon J, Walters R, Walsh N. Salivary immunoglobulinA response at rest and after exercise following a 48 h period of fluid and/or energy restriction. Br J Nutr. 2007;97:1109–16.
32. Walsh N, Gleeson M, Pyne D, Nieman D, Dhabhar F, Shephard R, et al. Position statement. Part two: maintaining immune health. Exerc Immunol Rev. 2011;17:64–103.
33. Bishop N, Gleeson M. Acute and chronic effects of exercise on markers of mucosal immunity. Front Biosci. 2009;14:4444–56.
34. Akimoto T, Kumai Y, Akama T, Hayashi E, Murakami H, Soma R, et al. Effects of 12 months of exercise training on salivary secretory IgA levels in elderly subjects. Br J Sports Med. 2003;37:76–9.
35. Klentrou P, Cieslak T, MacNeil M, Vintinner A, Plyley M. Effect of moderate exercise on salivary immunoglobulin A and infection risk in humans. Eur J Appl Physiol. 2002;87:153–8.
36. Moreira A, Mortatti AL, Arruda AFS, Freitas CG, de Arruda M, Aoki MS. Salivary IgA response and upper respiratory tract infection symptoms during a 21-week competitive season in young soccer players. J Strength Cond Res. 2014;28(2):467–73.
37. McKune A, Smith L, Semple S, Wadee A. Influence of ultra-endurance exercise on immunoglobulin isotypes and subclasses. Br J Sports Med. 2005;39:665–70.
38. Petibois C, Cazorla G, Deleris G. The biological and metabolic adaptations to 12 months training in elite rowers. Int J Sports Med. 2003;24:36–42.
39. Poortmans J. Serum protein determination during short exhaustive physical activity. J Appl Physiol. 1971;30:190–2.
40. Poortmans J, Haralambie G. Biochemical changes in a 100 km run: proteins in serum and urine. Eur J Phys. 1979;40:245–54.

41. Hejazi K, Hosseini S. Influence of selected exercise on serum immunoglobulin, testosterone and cortisol in semi-endurance elite runners. Asian J Sports Med. 2012;3:185–92.
42. Israel S, Buhl B, Neumann G. Die Konzentration der Immunglobuline A, G und M im Serum bei trainierten und untrainierten sowie nach verschiedenen sportlichen Ausdauerleistungen. Med Sport (Berlin). 1982;22:225–31.
43. Mashiko T, Umeda T, Nakaji S, Sugawara K. Effects of exercise on the physical condition of college rugby players during summer training camp. Br J Sports Med. 2004;38:186–90.
44. Nehlsen-Cannarella S, Nieman D, Jessen J, Chang L, Gusewitch G, Blix G, et al. The effects of acute moderate exercise on lymphocyte function and serum immunoglobulin levels. Int J Sports Med. 1991;12:391–8.
45. Nieman D, Nehlsen-Cannarella S. The effects of acute and chronic exercise on immunoglobulins. Sports Med. 1991;11:183–201.
46. Nielsen A, Pedersen B. The biological roles of exercise-induced cytokines: IL-6, IL-8, and IL-15. Appl Physiol Nutr Metab. 2007;32:833–9.
47. Nieman D, Dumke C, Henson D, McAnulty S, McAnulty L, Lind R, et al. Immune and oxidative changes during and following the Western states endurance run. Int J Sports Med. 2003;24:541–7.
48. Nieman D, Henson D, Smith L, Utter A, Vinci D, Davis J, et al. Cytokine changes after a marathon race. J Appl Physiol. 2001;91:109–14.
49. Pedersen B, Fischer C. Beneficial health effects of exercise--the role of IL-6 as a myokine. Trends Pharmacol Sci. 2007;28:152–6.
50. Cox A, Pyne D, Saunders P, Callister R, Gleeson M. Cytokine responses to treadmill running in healthy and illness-prone athletes. Med Sci Sports Exerc. 2007;39:1918–26.
51. Suzuki K, Nakaji S, Yamada M, Totsuka M, Sato K, Sugawara K. Systemic inflammatory response to exhaustive exercise. Cytokine kinetics. Exerc Immunol Rev. 2002;8:6–48.
52. Kurowski M, Jurczyk J, Moskwa S, Jarzębska M, Krysztofiak H, Kowalski ML. Winter ambient training conditions are associated with increased bronchial hyperreactivity and with shifts in serum innate immunity proteins in young competitive speed skaters. Arch Med Sci. 2018;14(1):60–8.
53. Gleeson M. Immune system adaptation in elite athletes. Curr Opin Clin Nutr Metab Care. 2006;9:659–65.
54. Lancaster G, Halson S, Khan Q, Drysdale P, Wallace F, Jeukendrup A, et al. Effects of acute exhaustive exercise and chronic exercise training on type 1 and type 2 T lymphocytes. Exerc Immunol Rev. 2004;10:91–106.
55. Kurowski M, Jurczyk J, Jarzębska M, Moskwa S, Makowska J, Krysztofiak H, et al. Association of serum Clara cell protein CC16 with respiratory infections and immune response to respiratory pathogens in elite athletes. Respir Res. 2014;15:45.
56. Peters EM, Goetzsche JM, Grobbelaar B, Noakes TD. Vitamin C supplementation reduces the incidence of postrace symptoms of upper-respiratory-tract infection in ultramarathon runners. Am J Clin Nutr. 1993;57:170–4.
57. Nieman DC. Exercise immunology: nutritional countermeasures. Can J Appl Physiol. 2001;26:S45–55.
58. Peck MD. Interactions of lipids with immune functions. Biochemical effects of dietary fatty lipids on plasma membranes. J Nutr Biochem. 1994;5:466–78.
59. Fraker PJ, Gershwin ME, Good RA, Prasad A. Interrelationships between zinc and immune function. Fed Proc. 1986;45:1474–9.
60. Barrett B, Vohmann M, Calabrese C. Echinacea for upper respiratory infection. J Fam Pract. 1999;48:628–35.
61. Moreira A, Kikkomen RA, Delgado L, Fonseca JA, Korpela R, Haahtela H. Nutritional modulation of exercise induced immunosuppression in athletes: a systematic review and metanalysis. Eur J Clin Nutr. 2007;61(4):443–60.
62. Constantini N, Ken-Dror A, Eliakim A, Galatzkia L, Morag A, Mann G, Volach B, Asharov J, Shoval D. Vaccinations in sports and recommendations for immunization against flu, hepatitis a and hepatitis B. Harefuah. 2001;1191-5(1228):140.

Part IV

Respiratory Training and Rehabilitation

Respiratory Muscle Training 10

Samuel Verges

Abstract

The respiratory muscles have a key role within the respiratory system allowing air to be pumped in and out the lungs. As other skeletal muscles, they can fatigue and be affected by pathological mechanisms leading to muscle weakness. They can also be trained using specific strategies inducing overloading and adaptations over time. Inspiratory muscle resistive training and respiratory muscle isocapnic hyperpnea training are the two main training methods used to increase respiratory muscle strength and endurance. It has been shown in healthy subjects that these kinds of training improve not only respiratory muscle function but also endurance exercise performance. In several diseases, patients can also benefit from respiratory muscle training which can improve respiratory function, exercise capacities, symptoms, and quality of life. As other training and rehabilitative methods, also respiratory muscle training, to be efficient, requires appropriate modalities, intensity, and duration as well as evaluations of the training effects.

10.1 Introduction

The respiratory muscles have a key physiological role within the respiratory system by allowing movement of air within the lungs to the alveoli. This is critical for gas exchange, especially for oxygen intake and carbon dioxide output. The respiratory muscles drive the ventilatory responses to various stimuli and in particular to physical exercise. An important concept to consider is the balance between the ventilatory demand (or respiratory load) and the respiratory muscle capacity (Fig. 10.1). Any disproportionate increase of the ventilatory demand or respiratory

S. Verges (✉)
Grenoble Alpes University and INSERM, Grenoble, France
e-mail: SVerges@chu-grenoble.fr

A. Cogo et al. (eds.), *Exercise and Sports Pulmonology*,
https://doi.org/10.1007/978-3-030-05258-4_10

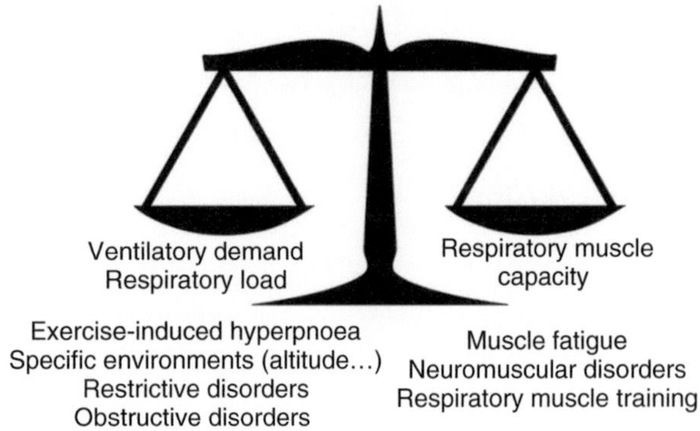

Fig. 10.1 The respiratory balance represented as the results of the respiratory load and the respiratory capacity. Some factors able to influence one side or the other of the balance are also indicated

load compared to the ability of the respiratory muscle output to accommodate this demand/load will result in an altered respiratory balance which will not be sustainable for a long period and which may induce deleterious consequences. An increase in ventilatory demand or respiratory load can be generated in healthy subjects by exercise (and the associated hyperpnea) and by exposure to specific environment such as altitude (and the associated hyperventilatory response). Several diseases can induce an increase in respiratory load, for instance, obstructive and restrictive respiratory diseases, which can enhance the airway resistance and reduce thoracic compliance. The capacity of the respiratory muscles to pump air in and out the lungs can be altered due to transitory (fatigue) or permanent (weakness) reduction in their ability to produce inspiratory and expiratory pressures. This can happen in healthy subjects following fatiguing physical exercise and in several disease conditions affecting the function (structure, metabolism, etc.) of the respiratory muscles and/or overloading the respiratory muscles. The capacity of the respiratory muscles to generate pressure and flow can also be enhanced by using specific training strategies (respiratory muscle training).

10.2 General Concepts

The respiratory muscles should be considered as any other skeletal muscles on many aspects. Muscles have two functions, i.e., to develop force and to shorten. In the respiratory system, force is expressed as changes in pressures, while shortening is expressed as changes in lung volume or displacement of the chest wall. Therefore, the respiratory muscle function is generally performed by measuring the respiratory pressures and lung volumes. Respiratory muscle strength in clinical setting is usually assessed based on maximal voluntary inspiratory or expiratory maneuvers against an occluded airway, leading to the measurement of maximal inspiratory (PI_{max}) or expiratory (PE_{max}) pressures. Additional clinical measurements such as sniff tests and peak expiratory cough flow can also be used. In research setting, electrical or magnetic stimulation of the motoneurons innervating the respiratory muscles (the phrenic nerves for the diaphragm, for instance) can be used to provide a non-volitional and more specific assessment of respiratory muscle force production capacities. Respiratory muscle endurance can also be evaluated by asking the subjects to sustain an increased respiratory load over a prolonged period, e.g., by breathing with an added inspiratory resistance or by performing isocapnic hyperpnea. The time to task failure (i.e., the inability to sustain the respiratory load) is generally the index used to quantify respiratory muscle endurance. The reader is referred to the American Thoracic Society/European Respiratory Society position statement on respiratory muscle testing [1] for more information on this topic.

Based on careful respiratory muscle strength measurements, respiratory muscle fatigue or weakness can be demonstrated. Muscle fatigue is defined as a transient reduction in the ability to generate force (or pressure for the respiratory muscles) following a period of overload (e.g., exercise-induced hyperpnea) followed by progressive recovery of the initial muscle abilities. Muscle weakness is defined as a permanent reduction of the ability to generate strength compared to the expected muscle force capacities. Respiratory muscle fatigue has been demonstrated in healthy subjects, for instance, following intense whole-body exercise (e.g., cycling) by showing reduced maximal voluntary or evoked (by electrical or magnetic stimulations) inspiratory and expiratory muscle pressures after, compared to before exercise [2–5]. Respiratory muscle weakness can be observed in various diseases by comparing PI_{max} and PE_{max}, for instance, to theoretical values that provide expected values for a healthy subject depending on sex, age, and body size.

As other skeletal muscles, respiratory muscles can adapt to repetitive increase in load; in fact, they can increase their capacities in response to specific training program. Therefore, the key principles of physical training apply also to respiratory muscle training. To induce muscle adaptations, training programs should induce overloading, i.e., a significant increase in the total work the muscles have to perform over a prolonged period of time. The training load will depend on the intensity of the work during the exercise training session (e.g., the strength and speed the muscle has to develop and the number of contractions) and the number of training sessions per week. If the intensity of the training session is too low, or the number of training sessions per week is insufficient, the training will not induce significant muscle improvement. Similarly, if the exercise training intensity is too high, or the training sessions are too frequent and do not allow proper recovery, the training will also fail in improving muscle capacities. Therefore, the efficiency of respiratory muscle training will depend on the determination of the appropriate training program and should be evaluated by objective methods such as the assessment of respiratory muscle strength and endurance. In addition to improving respiratory muscle strength and endurance, the ultimate goal of respiratory muscle training is to increase performance of sportsmen and to improve health status, functional capacities, and quality of life of patients. Hence, in addition to the effects of respiratory muscle training on respiratory muscle function (respiratory muscle strength and endurance, for instance), these important outcomes (e.g., exercise endurance, respiratory symptoms, health-related quality of life, etc.) also have to be evaluated to objectivize the potential effect of this kind of training.

10.3 Respiratory Muscle Training in Healthy Individuals

Two main methods of respiratory muscle training have been developed and evaluated by specific studies: inspiratory muscle resistive training and respiratory muscle isocapnic hyperpnea training.

Because the inspiratory muscles are mainly involved in breathing, they have been targeted by training methods consisting in repetitive contractions against a load, either a resistance (which has the disadvantage to be flow-dependent) or more usually a threshold-loading valve, which opens only when the subject produces a certain negative pressure during inspiration. The load during inspiratory resistive training is set based on the target mouth inspiratory pressure to produce, expressed as a percentage of the PI_{max} and on the number of contractions (inspiratory maneuvers) to perform per session. It is also important to consider the lung volume and the respiratory pattern during inspiratory resistive training, because these parameters will influence the length and the speed of the inspiratory muscle contractions and therefore the muscle adaptations induced by training. Depending on studies, inspiratory muscle pressure during inspiratory muscle training varies between 30% PI_{max} (low-intensity training) and 80% PI_{max} (high-intensity training).

In order to mimic heavy breathing as during intense physical exercise, respiratory muscle training methods based on isocapnic voluntary hyperpnea have been developed. During this kind of training, subjects have to increase their ventilation to >60% of their maximal minute ventilation (measured over 10–15 s or calculated theoretically as FEV1 × 35–40) and to sustain this high level of ventilation for 20–30 min per session. With this method, both inspiratory and expiratory muscles are recruited and trained. In order to prevent hypocapnia, induced by hyperventilation, the use of a specific device allowing partial rebreathing is required to maintain a stable CO_2 (isocapnic hyperpnea). The device should also provide feedback on the appropriate tidal volume and breathing frequency to sustain during the training session. As for inspiratory resistive breathing, several devices are commercially available and allow proper training conditions (e.g., POWERbreathe© K-Series for inspiratory resistive training, Spirotiger© for isocapnic hyperpnea training). Other devices do not provide proper feedback and may therefore prevent the subjects to benefit from training, as expected.

Both inspiratory resistive training and isocapnic hyperpnea training programs over 4–8 weeks (five to seven sessions per week) have been shown to improve significantly respiratory muscle function in healthy subjects. Inspiratory resistive training mostly improves inspiratory muscle strength (as shown by increased PI_{max} [6]), while isocapnic hyperpnea training mainly increases respiratory muscle endurance (as demonstrated by an increase in time to exhaustion during a hyperpnea endurance test [7]) (Fig. 10.2). In addition to increasing respiratory muscle strength and endurance,

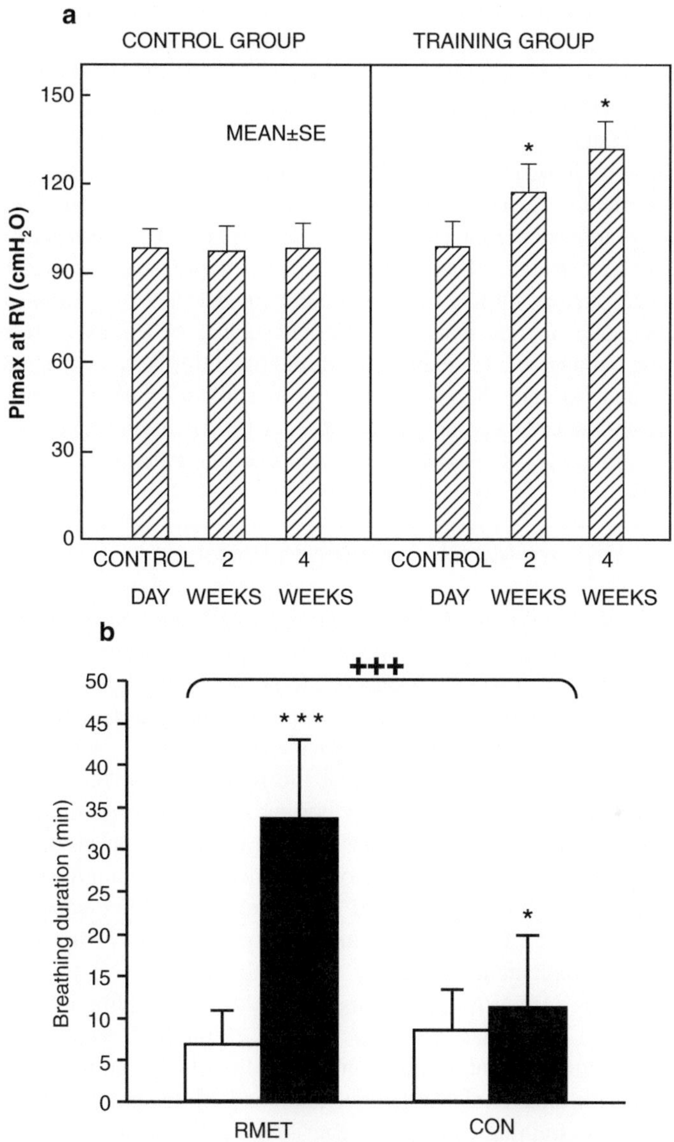

Fig. 10.2 (**a**) The effect of 4 weeks of inspiratory resistive training on maximal inspiratory pressure (PI_{max}, measured at residual volume, RV) [6]; and (**b**) of 4–8 weeks of respiratory muscle isocapnic hyperpnea training (RMET) on breathing endurance [7]. Changes are provided for a training group and for a control (CON) group

respiratory muscle training has been shown to increase performance during various types of exercise in healthy subjects (cycling, running, rowing, etc.). Maximal oxygen consumption and maximal aerobic power output generally did not improve after respiratory muscle training, while submaximal endurance performance (time to exhaustion at submaximal intensity, distance during a time trial, time for a given distance) has been shown to increase in several studies. A systematic review and meta-analysis [8] have assessed the effect of respiratory muscle training on exercise endurance performance and the factors influencing the change in exercise endurance performance in healthy subjects. The conclusions indicated (1) a significant improvement in exercise performance after respiratory muscle training, which was detected by constant load tests, time trials but not by incremental tests; (2) less fit subjects benefit the most from respiratory muscle training; and (3) improvements did not differ between inspiratory resistive training and isocapnic hyperpnea training. Respiratory muscle training can also reduce exercise-induced respiratory muscle fatigue [9, 10], improve respiratory sensations during exercise [7], and reduce blood lactate concentration [11, 12].

10.4 Respiratory Muscle Training in Patients

Respiratory muscle training has been tested in several neuromuscular disorders as an adjunct to other rehabilitation strategies. In patients with multiple sclerosis and lateral amyotrophic sclerosis, for instance, it improves lung function and respiratory muscle strength [13]. In spinal cord injury, respiratory muscle training is recognized as an efficient method to improve vital capacity, PI_{max} and PE_{max}, while the potential positive effects on dyspnea and quality of life remain to be confirmed [14]. The effect of respiratory muscle training in chronic obstructive pulmonary diseases (COPD) has been assessed in numerous clinical trials. Recent systematic review and meta-analysis indicate that inspiratory resistive training improves inspiratory muscle strength, exercise capacity, and quality of life and decreases dyspnea in COPD patients [15, 16]. In asthmatic patients, respiratory muscle training has been proposed as a complementary therapy to the pharmacological treatments. This intervention seems to improve symptoms and quality of life, although the evidence remains modest [17]. Similarly, in cystic fibrosis, respiratory muscle training may contribute to a better respiratory function and quality of life, but further studies are required to provide clear clinical recommendations [18]. In cardiovascular diseases, respiratory muscle training can also be considered as an attractive intervention in addition to standard care. Adding inspiratory resistive training to whole-body exercise training program induces significant improvement in inspiratory muscle strength and quality of life of patients with heart failure, for instance [19]. Respiratory muscle training should also be considered in obese patients [20, 21] and after stroke [22, 23]. There are recent evidences that respiratory muscle training is feasible and safe in ventilated patients and that this kind of intervention can reduce the weaning period and improve weaning success rates [24]. Respiratory muscle training can be used pre- and post-thoracic and upper-abdominal surgery in order to improve lung function and to reduce postoperative respiratory complications [25, 26].

10.5 Recommendations and Remaining Questions

Respiratory muscle training has been proposed as a complementary intervention in the management of a range of diseases. It appears to consistently improve respiratory muscle function and has been associated in several diseases with a clinically important improvement in symptoms such as dyspnea, impaired cough, functional capacity (such as exercise performance), and quality of life. However, its indications and potential contraindications, the type of evaluations, the optimal training modalities (intensity, duration and number of sessions, etc.), and the selection of patients who may benefit the most remain to be better clarified. It seems that patients with respiratory muscle weakness and respiratory symptoms (dyspnea, impaired cough, etc.) could be good candidates.

Inspiratory resistive training should be done with a flow-independent device (by using a threshold-loading valve, for instance) and a controlled pattern of breathing ensuring the patient inspires a sufficient volume of air at an appropriate rate. Hyperpnea training should be done with a device ensuring isocapnia (by partial rebreathing, for instance) and providing feedback to the patients regarding tidal volume and breathing frequency. A sufficient intensity has to be determined for the first training sessions (see Table 10.1). Then, it is critical to increase the intensity at least weekly as the patient's respiratory muscle capacity will progressively increase in order to maintain an appropriate muscle overloading. A sufficient number of sessions per week are also critical (see Table 10.1), and a minimum training program of 4 weeks appears reasonable. Posture and breathing technique, especially if the respiratory muscle training session is supervised by a professional health care, should be optimized since respiratory muscle training is also able to influence and improve these aspects. As for any kind of exercise training, the benefits of respiratory muscle training will progressively disappear after the end of the training program if the overloading of the muscle (i.e., the training load) is not present anymore. Whether inspiratory resistive training and isocapnic hyperpnea training provide

Table 10.1 Respiratory muscle training modalities

	Inspiratory resistive training	Isocapnic hyperpnea training
Description	Repetitive inspiratory maneuvers against a fixed resistance similar to strength training	Heavy breathing without resistance but at high minute ventilation mimicking breathing during intense exercise, similar to muscle endurance training. Partial rebreathing allows isocapnia
Session duration	10–15 min, twice a day	20–30 min
Frequency	Five to seven sessions per week	Five sessions per week
Intensity	30–50% PI_{max} (adjusted weekly)	50–60% of maximum minute ventilation (adjusted weekly), tidal volume close to 50% vital capacity
Device feedback	Information on breathing pattern and inspired volume are important	Information on tidal volume and breathing frequency are important

similar benefits or should be selected for certain types of patients remain unknown. Specific expiratory muscle resistive training may also be considered in some conditions [27].

Key Points

- Respiratory muscle weakness and fatigue have been identified in a variety of conditions.
- They can be responsible for significant functional impairments, symptoms, and impaired quality of life of patients.
- Respiratory muscles can be trained similarly to other skeletal muscles to increase their strength and endurance.
- Respiratory muscle training by either inspiratory resistive breathing or isocapnic hyperpnea induces significant improvement in respiratory muscle function in healthy subjects and several different diseases.
- These increases in respiratory muscle function can be associated with enhanced exercise performance and respiratory function and improvements in symptoms and quality of life.

References

1. ATS/ERS. ATS/ERS statement on respiratory muscle testing. Am J Respir Crit Care Med. 2002;166(4):518–624.
2. Johnson BD, Babcock MA, Suman OE, Dempsey JA. Exercise-induced diaphragmatic fatigue in healthy humans. J Physiol. 1993;460:385–405.
3. Taylor BJ, How SC, Romer LM. Exercise-induced abdominal muscle fatigue in healthy humans. J Appl Physiol. 2006;100(5):1554–62.
4. Verges S, Notter D, Spengler CM. Influence of diaphragm and rib cage muscle fatigue on breathing during endurance exercise. Respir Physiol Neurobiol. 2006;154(3):431–42.
5. Verges S, Schulz C, Perret C, Spengler CM. Impaired abdominal muscle contractility after high-intensity exhaustive exercise assessed by magnetic stimulation. Muscle Nerve. 2006;34(4):423–30.
6. Suzuki S, Yoshiike Y, Suzuki M, Akahori T, Hasegawa A, Okubo T. Inspiratory muscle training and respiratory sensation during treadmill exercise. Chest. 1993;104(1):197–202.
7. Verges S, Boutellier U, Spengler CM. Effect of respiratory muscle endurance training on respiratory sensations, respiratory control and exercise performance: a 15-year experience. Respir Physiol Neurobiol. 2008;161(1):16–22.
8. Illi SK, Held U, Frank I, Spengler CM. Effect of respiratory muscle training on exercise performance in healthy individuals: a systematic review and meta-analysis. Sports Med. 2012;42(8):707–24.
9. Romer LM, McConnell AK, Jones DA. Inspiratory muscle fatigue in trained cyclists: effects of inspiratory muscle training. Med Sci Sports Exerc. 2002;34(5):785–92.
10. Verges S, Lenherr O, Haner AC, Schulz C, Spengler CM. Increased fatigue resistance of respiratory muscles during exercise after respiratory muscle endurance training. Am J Physiol Regul Integr Comp Physiol. 2007;292(3):R1246–53.
11. McConnell AK, Sharpe GR. The effect of inspiratory muscle training upon maximum lactate steady-state and blood lactate concentration. Eur J Appl Physiol. 2005;94(3):277–84.

12. Spengler CM, Roos M, Laube SM, Boutellier U. Decreased exercise blood lactate concentrations after respiratory endurance training in humans. Eur J Appl Physiol Occup Physiol. 1999;79(4):299–305.
13. Ferreira G, Costa A, Plentz R, Coronel C, Sbruzzi G. Respiratory training improved ventilatory function and respiratory muscle strength in patients with multiple sclerosis and lateral amyotrophic sclerosis: systematic review and meta-analysis. Physiotherapy. 2016;102(3):221–8.
14. Berlowitz D, Tamplin J. Respiratory muscle training for cervical spinal cord injury. Cochrane Database Syst Rev. 2013;23(7):CD008507.
15. Beaumont M, Forget P, Couturaud F, Reychler G. Effects of inspiratory muscle training in COPD patients: a systematic review and meta-analysis. Clin Respir J. 2018;12:2178.
16. Gosselink R, De Vos J, van den Heuvel SP, Segers J, Decramer M, Kwakkel G. Impact of inspiratory muscle training in patients with COPD: what is the evidence? Eur Respir J. 2011;37(2):416–25.
17. Freitas D, Holloway E, Bruno S, Chaves G, Fregonezi G, Mendonça K. Breathing exercises for adults with asthma. Cochrane Database Syst Rev. 2013;1(10):CD001277.
18. Houston BW, Mills N, Solis-Moya A. Inspiratory muscle training for cystic fibrosis. Cochrane Database Syst Rev. 2013;21(11):CD006112.
19. Neto M, Martinez B, Conceiçao C, Carvalho V. Combined exercise and inspiratory muscle training in patients with heart failure: a SYSTEMATIC REVIEW AND META-ANALYSIS. J Cardiopulm Rehabil Prev. 2016;36(6):395–401.
20. Villiot-Danger J, Villiot-Danger E, Borel J, Pépin J, Wuyam B, Verges S. Respiratory muscle endurance training in obese patients. Int J Obes. 2011;35(5):692–9.
21. Vivodtzev I, Tamisier R, Croteau M, Borel J, Grangier A, Wuyam B, et al. Ventilatory support or respiratory muscle training as adjuncts to exercise in obese CPAP-treated patients with obstructive sleep apnoea: a randomised controlled trial. Thorax. 2018;73:634.
22. Martin-Valero R, De La Casa Almeida M, Casuso-Holgado M, Heredia-Madrazo A. Systematic review of inspiratory muscle training after cerebrovascular accident. Respir Care. 2015;60(11):1652–9.
23. Menezes K, Nascimento L, Ada L, Polese J, Avelino P, Teixeira-Salmena L. Respiratory muscle training increases respiratory muscle strength and reduces respiratory complications after stroke: a systematic review. J Physiother. 2016;62(3):138–44.
24. Bissett B, Leditschke I, Paratz J, Boots R. Respiratory dysfunction in ventilated patients: can inspiratory muscle training help? Anaesth Intensive Care. 2012;40(2):236–46.
25. Katsura M, Kuriyama A, Takeshima T, Fukuhara S, Furukawa T. Preoperative inspiratory muscle training for postoperative pulmonary complications in adults undergoing cardiac and major abdominal surgery. Cochrane Database Syst Rev. 2015;5(10):CD010356.
26. Mans C, Reeve J, Elkins M. Postoperative outcomes following preoperative inspiratory muscle training in patients undergoing cardiothoracic or upper abdominal surgery: a systematic review and meta analysis. Clin Rehabil. 2015;29(5):426–38.
27. Laciuga H, Rosenbek J, Davenport P, Sapienza C. Functional outcomes associated with expiratory muscle strength training: narrative review. J Rehabil Res Dev. 2014;51(4):535–46.

Pulmonary Rehabilitation

11

Francesca de Blasio, Francesco de Blasio, and Enrico Clini

Abstract

Disability in chronic respiratory diseases (CRD) represents the impact of the disease on the patient's life. Chronic airway diseases, included but not limited to COPD, are leading this burden.

Overall, the mobility-related dyspnea and the resulting decrease in exercise capacity substantially contribute to increased risk of disability, even after taking lung function impairment into account. Therefore, non-pharmacological interventions such as pulmonary rehabilitation (PR) might be particularly beneficial for these symptomatic patients to limit and to counteract the progressive loss of physical function and related problems.

In this chapter we will discuss the most recent evidence related to the assessment of individual's disability in this population, and we will describe the variety of methods used in the clinical process of care called PR.

To date, PR results in substantial effectiveness when applied at the very early onset of disability in individuals suffering from CRD. Programme composition and strategies aimed at behavioural changes in the long-term appear the keys for success in the clinical practice.

F. de Blasio
Department of Medical Science, University of Turin, Torino, TO, Italy

F. de Blasio
Department of Medicine and Health Sciences, University of Molise, Campobasso, CB, Italy

Respiratory Medicine and Pulmonary Rehabilitation Section, Clinic Center S.p.A. Private Hospital, Naples, Italy

E. Clini (✉)
Department of Medical and Surgical Sciences, University of Modena Reggio Emilia, Modena, MO, Italy

Pneumology Unit, University Hospital of Modena, Modena, MO, Italy
e-mail: enrico.clini@unimore.it

A. Cogo et al. (eds.), *Exercise and Sports Pulmonology*,
https://doi.org/10.1007/978-3-030-05258-4_11

11.1 Disability in Chronic Respiratory Disease

11.1.1 The Vicious Cycle of Dyspnea

Disability in chronic respiratory diseases (CRD) is a significant health burden with relevant implications both for the individual patient and the society: indeed, it represents the real impact of the disease on the patient's life, with an important influence on the society overall work productivity [1]. Chronic obstructive pulmonary disease (COPD), among these, accounts for one of the top five causes of disability all over the world. Although respiratory impairment contributes to and increases the risk of disability, the presence of limitations of the individual's general function and the occurrence of non-respiratory symptoms with the presence of comorbid extra-pulmonary conditions have a great impact on disablement as well. Table 11.1 shows some of the conditions that may cause or aggravate dyspnea in individuals with CRD. The mobility-related dyspnea and the resulting decrease in exercise capacity substantially contribute to increased risk of disability, even after taking lung function impairment into account [2]. Therefore, the assessment and treatment of airway obstruction, as for patients with COPD, are not sufficient to prevent and care for the development of individual's disability.

11.1.2 Peripheral Muscle Weakness as the Hallmark of Disease

Limb muscle dysfunction, defined as the reduction of either strength or endurance (or both) [3], is frequent in patients with CRD [2], and in particular in COPD, with muscle fibre shift, atrophy and changes in capillarization that are commonly seen in their peripheral muscles [4, 5].

Although the extent of muscle atrophy and weakness is greater in advanced disease, it is important to recognize that muscle dysfunction may even occur at an early stage [5, 6]. For instance, symptomatic COPD patients referred to and entering

Table 11.1 Causes of dyspnea in patients with chronic respiratory disease

Increased resistive work of breathing from airflow limitation
Increased elastic work of breathing and "pseudo-restriction" from static and dynamic hyperinflation
Physical and cardiovascular deconditioning from sedentary
Gas exchange abnormalities: hypoxemia and increased physiologic dead space
Cardiovascular limitations: cardiac or peripheral vascular co-morbidity, leading to early lactate production with exercise
Skeletal muscle abnormalities: decreased mass, fibre-type alterations (reduction in type I, increase in type IIx), capillarization defect, decreased oxidative enzymes, also leading to early lactate production with exercise
Coexisting obesity, increasing workload requirements for a specific task
Anxiety associated with dyspnea-producing activity

a rehabilitation programme have already lost about 30% of their muscle mass and strength [7].

The prevalence of peripheral muscle weakness varies (from 20 to 40%) among patients, with a typical interindividual heterogeneity, but increases with the severity of the respiratory condition.

Furthermore, peripheral muscle weakness is not equally distributed among muscle groups: compared to the lower limbs, the strength of the upper limb muscles (although reduced) seems better preserved, especially in COPD [8], probably reflecting the heterogeneous distribution of muscle structural abnormalities. The annual rate decline in quadriceps strength in patients with COPD is 4.3% per year [9], in comparison with 1–2% per year in the elderly people. Although quadriceps muscles represent a typical example of a primary locomotor muscle that is underused in symptomatic patients who become sedentary, upper limb muscle function is also affected, as shown by a reduced handgrip strength during acute care [10].

Thus, muscle strength/weakness represents a clinical hallmark of several CRD and drives both individual's physical activity and functional capacity [11–13]. Weakness, in particular, has been associated with relevant negative outcomes such as dyspnea burden, exercise intolerance [14], morbidity, mortality [15, 16] and poor quality of life [9].

11.2 Assessment of Individual's Disability

11.2.1 Muscle Function

The assessment of muscle function (strength and endurance) is muscle group specific. It also varies depending on the measurement technique (isokinetic, isometric or isotonic) and the device used, which must be chosen based on their advantages and limits as well as on the desired information [16]. Some of the daily life activities rely on of the isometric contraction (e.g. carrying grocery bags, standing up from and sitting down on a chair, pushing and pulling). However, most of the functional activities of daily living may be better assessed by dynamic techniques, i.e. isokinetic (fixed speed of movement) and isotonic (fixed resistance applied to the muscle during the movement), which provide information on limb muscle function throughout the full range of motion at different speeds. Table 11.2 shows an overview of the methods used for assessing muscle function, as valid in COPD.

Muscle atrophy is another common manifestation of CRD and in COPD in particular. Atrophy can be included under the umbrella term of muscle dysfunction since the loss of muscle mass may have important implications on strength [7, 14, 17] and exercise tolerance [18–20]. Muscle atrophy is the main cause of weight loss in COPD patients [18] independently on the degree of airway obstruction [21], and it is a predictor of health status [22] and survival [23].

Several techniques are available to assess the mass of peripheral muscles [4]: anthropometry (mid-arm muscle circumference), bioelectric impedance analysis (BIA), dual-energy X-ray absorptiometry (DXA) as well as more advanced imaging

Table 11.2 Methods used to assess muscle function in patients with COPD

Measuring equipment	1. Strain-gauge or hand-held systems	2. Computerized dynamometers	3. Repetition maximum	4. Handgrip gauges	5. Non-volitional (electrical or magnetic nerve stimulation)
Type of contraction	Isometric	Isometric and isokinetic	Isotonic	Isometric	Isometric
Positioning of tested limbs in COPD studies [85–87]	Knee and hip flexed at 90°, elbow and/or shoulder flexed at 90°	*Isometric*: Elbow flexed at 90° knee flexed 60°–120°. *Isokinetic*: Knee flexed at 90° to full knee extension	Not well described in COPD studies	Elbow flexed at 90° with arm unsupported	Knee and hip flexed at 90°
Measurements of *muscle strength* in COPD studies [85, 88]	Perform 3 MVC, each lasting 3 to 5 s with at least 30-s rest between attempts. Highest value used as measurement of strength	Perform 3–5 MVC *(isometric: hold each contraction for 3 s)* and 1–2-min rest between attempts. Highest value used as measurement of strength	Not well described in COPD studies. Follow ACSM recommendations + use metronome and control start and stop positions	Perform 3 MVC with each lasting at least 3 s and 30-s rest between attempts. Highest value used as measurement of strength	Three stimulations at 100% stimulator output in a relaxed muscle. Highest value used as measurement of strength
Measurements of *muscle endurance* in COPD studies [87, 89–91]	Measure the time during which a contraction at 60–80% of MVC can be maintained in one set (only performed using strain gauges in COPD studies)	*Isometric*: Has not previously been used in COPD. *Isokinetic*: Measure the total amount of work performed from 1 set of 30 MVC at 60° or 90°/s	Number of repetitions performed. One set of dynamic contractions at 30–40% of MVC until exhaustion. Contraction cycle of 6–12 contractions	Has not previously been done in COPD. Strain-gauge procedure might be feasible	One set of repeated magnetic stimulations using a train of impulsions at 20–30% of MVC until a 70–80% reduction in force is seen

Advantages [88, 92–95]	Results are valid, reliable and reproducible. Easy to use, portable, time efficient and inexpensive	Results are valid, reliable and reproducible. Easy to standardize. Different speeds and angles could be tested	Assesses muscle function in the whole range of motion. Can be executed using available equipment	No familiarization, easy to use, easy to standardize	Less affected by external factors. Lower day-to-day variability than assessments of MVC
Limitations [85, 96–98]	Measures only in one angle. Standardization is crucial for validity and reliability	Low availability. Requires expensive equipment. Needs familiarization session	Time-consuming and more difficult to standardize than isometric measurements	Measures only handgrip strength	More in the realm of research. Requires expensive equipment

technologies like computed tomography (CT) and nuclear magnetic resonance (NMR). Two common methods to estimate muscle mass in clinical practice are BIA and DXA. BIA is a valid, non-invasive, inexpensive, quick and easy to perform technique that, like DXA, requires no active collaboration from the patient. Like most body composition methods (included DXA), BIA does not directly measure muscle mass but provides indirect estimates of fat-free mass (as a proxy of muscle mass) from the measurement of resistance of body tissues to an electric current passing through the body. Alternatively, directly measured raw BIA variables, such as phase angle, have been demonstrated to relate to muscle function, disease severity and prognosis in COPD patients better than fat-free mass estimates [24–27]. DXA is another valid, reliable, safe and non-invasive technique for assessment of muscle mass. It is based on the comparison of X-ray attenuations of two different energies measuring body composition with a higher degree of accuracy [28]. It is however more expensive and often less easily accessible than BIA in clinical settings [25].

11.2.2 Symptoms

As part of a comprehensive assessment of individual disability, quantifying symptoms (dyspnea and fatigue) through specific tools is crucial in order to describe the level of chronic disability and to retest changes following interventions (i.e. rehabilitation). Currently, a number of scales are available to classify symptoms, but the most widely used ones are the Modified British Medical Research Council (mMRC) and the Borg scale. The former is a simple measure of breathlessness, as the person perceives it. It ranges from 0 ("I only get breathless with strenuous exercise") to 4 ("I am too breathless to leave the house, or I am breathless when dressing"), and it is considered adequate for assessment of symptoms since it correlates well with health status [29] and mortality risk [30]. The latter measures the perception of symptoms (dyspnea or fatigue) during physical activity [31].

In addition to these "categorical", different scales (e.g. visual analogue scale, VAS) where the determination of the severity of dyspnea is of an analogical type can be used [32]. Finally, as part of the individual's overall health status, symptoms can be measured by generic or disease-specific questionnaires, such as the Chronic Respiratory Questionnaire (CRQ), the St. George Respiratory Questionnaire (SGRQ), the COPD Assessment Test (CAT) and the COPD Control Questionnaire (CCQ).

11.2.3 Exercise Capacity

Assessment of exercise capacity in patients with CRD can be obtained with a number of different methods that essentially can be divided into *field* or *laboratory* tests [33].

Field tests, such as the timed walk tests, are the most popular ones, because they are easy to perform and related to the individual's daily functional activities. On the other hand, they are performed at submaximal capacity and are not able to provide physiologic information about those complex mechanisms that may limit exercise on an individual basis.

Laboratory tests to physiologically assess the cardiopulmonary adaptation to exercise include the incremental (iCPET) and constant (cCPET) work rate cardiopulmonary exercise tests [34].

Cardiopulmonary exercise testing (CPET)—CPET represents the gold standard for exercise performance assessment [34]. Indeed, continuous displacement of cardiovascular, respiratory and haematological parameters as well as the individual's perception (symptoms) during exercise provides information about the physiological reserve, systems' interaction and mechanisms of limitation to exercise. In particular, the typical iCPET provides continuous data on the ventilatory adaptation (e.g. tidal volume (VT) and minute ventilation (VE)), respiratory gas exchange (e.g. oxygen saturation ($SatO_2$), oxygen uptake (VO_2) and carbon dioxide output (VCO_2)), cardiovascular response (e.g. cardiac frequency (HR), blood pressure (BP) and cardiac rhythm) and symptom response (e.g. perceived dyspnea and/or leg fatigue reported by a numeric or visual analogue scale). Given these characteristics, CPET enables accurate determination of the physiologic reserves of the heart and lungs as well as functional capacity [35, 36], and it is a test therefore used both to assess the individual's normality and to look for cardiopulmonary limitations. Notwithstanding, although its use is becoming more widespread, CPET still remains largely underutilized in the general practice due to costs for the appropriate setting and apparatus, as well as for complexity and limited practicability in the more severe diseases. An example of CPET laboratory setting is shown in Fig. 11.1.

Submaximal field tests and physical functioning–Standard submaximal exercise tests to estimate maximal oxygen uptake (as described in the *American College of Sports Medicine's Guidelines for Exercise Testing and Prescription*) [37] are based on the primary assumption that the maximal heart rate of the individual undergoing this test is similar to a predicted maximal heart rate based on a formula such as "220 minus age". Such formulae may be applied with caution to healthy individuals as long as one is aware of the significant interindividual variability (SD = 10–12 beats/min) of his/her maximal heart rate. However, many studies that measured maximal aerobic capacity of persons with a variety of medical conditions such as cardiovascular, metabolic, neurologic or neuromuscular disease found significantly lower maximal heart rates in these patient populations.

Compared with maximal exercise testing, submaximal exercise testing appears to have greater applicability to the "world" of healthcare practitioners (physicians, physical therapists, nurses) in their role as clinical exercise specialists. Therefore, these tests are usually the preferred choice for the majority of individuals suffering from CRD that are likely to be limited by dyspnea and/or fatigue or also present abnormal gait and impaired balance.

There are several submaximal tests validated for the clinical practice as described in Table 11.3.

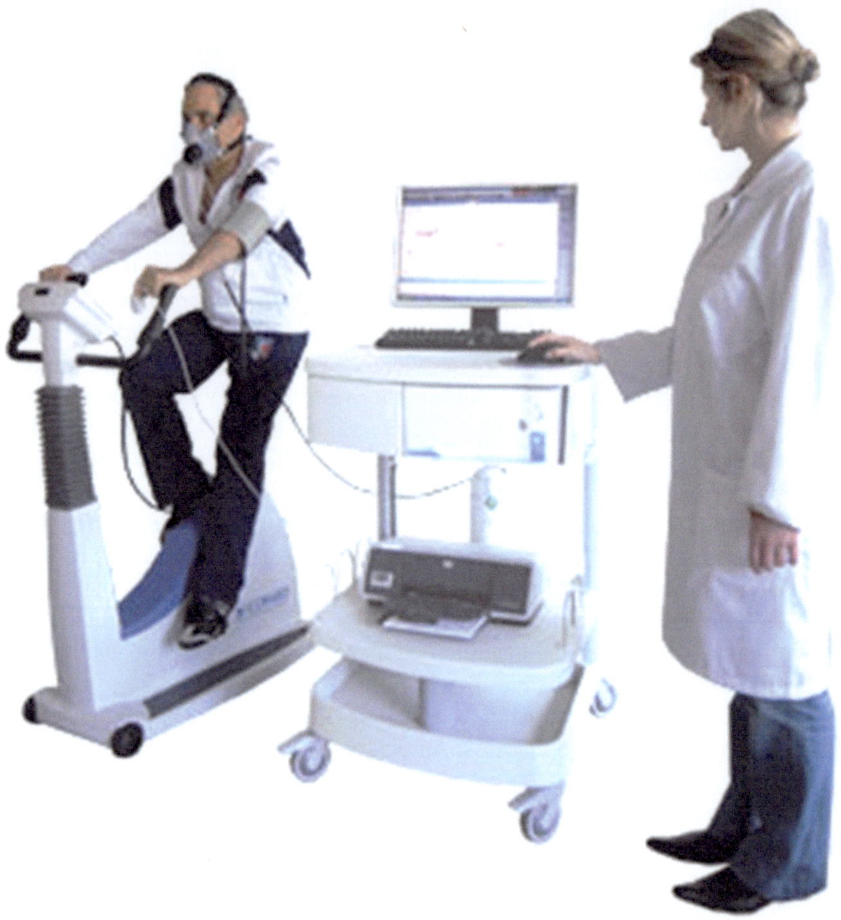

Fig. 11.1 Example of a cardiopulmonary exercise testing in a laboratory setting

In the clinical practice, however, 6-min walk distance (6 MWD), shuttle walk (SW), timed up and go (TUG) and sit to stand (STS) are those more frequently used in patients suffering from CRD. Therefore, we briefly describe each of these four as follows:

- The *6 MWT* is the easiest test that requires a 100 ft. hallway. It measures the distance that a patient can quickly walk on a flat, hard surface in a period of 6 min (the 6 MWD) [38]. Before the test starts, the patient should sit at rest in a chair, located near the starting position, for at least 10 min. During this time, pulse and blood pressure will be measured, and patient's baseline dyspnea and overall fatigue will be recorded using the Borg scale [39]. Patient will be instructed to walk back and forth in the hallway for 6 min. At the end of the test, post-walk Borg dyspnea and fatigue levels will be recorded again.

Table 11.3 Submaximal exercise capacity assessment tests

Tests	Aims	Administration	Main outcomes	Advantages	Limitations
Constant-rate cardiopulmonary test	Endurance cardiorespiratory exercise capacity	Constant work rate proportional to peak exercise capacity (e.g. 60% of peak work rate), until exhaustiontreadmill or bicycle	Time until exhaustioncardiorespiratory function variables	Greater sensitivity to identify changes after interventioncardiopulmonary function diagnosticcardiovascular risk assessment	Requires a previous maximal test equipment and certified personnel-related costs
6-min walk test	Functional exercise capacity	Walking back and forth on a 30 m courseself-paced speed	Total distance walked in 6 min	Reliability and validitylow complexity and costgood correlation with activities of daily living	Does not provide cardiopulmonary diagnosisno detailed information on physiological variables ($V^{\cdot}O_2$, $V^{\cdot}E$) and exercise limitation mechanisms may require a previous familiarization test
Incremental shuttle walk test	Functional exercise capacity	Walking back and forth on a 10 m course with paced increments of walking speed, until inability to keep the pace	Total distance walked until exhaustion	Fast to prepare and performgood correlation with $V^{\cdot}O2$ max in CPETlow cost	Higher risk of cardiovascular eventsless widespread usesubject to patient motivation
Endurance shuttle walk test	Endurance functional exercise capacity	Walking back and forth on a 10 m course with fixed paced of walking speed, until inability to keep the pace	Time until exhaustion	Good correlation with CPET cardiorespiratory responsereliability, validity, responsivenesslow complexity and cost	Requires a previous incremental shuttle testsame as incremental shuttle test

- The *SW* test is similar to the 6MWT, but it uses a series of audio signals to direct the walking pace [40, 41]. During this test, patients will be asked to walk between two cones spaced 10 m apart. Patients will start by walking at a very slow pace; this pace is set by a *beep*. Patients will walk around the 10 m course and will turn around a cone at the first beep and around the second cone at the next beep. The beeps will gradually get faster, which means patients will start to walk at a quicker pace, getting faster until he/she cannot keep up with the set pace, or until he/she is too tired or too breathless to continue.
- The *TUG* test is performed using a standard chair (height of the seat being 45 cm). Subjects are seated with their back supported against the chair. They are instructed to stand up, walk 3 m to a mark on the floor, cross the mark, turn around, walk back to the chair and sit down. The task needs to be performed at their normal comfortable pace. A stopwatch is started on the word "go" and stopped as the subject sit down; the time recorded in seconds represents the outcome value. Applicability and repeatability of this test in patients with COPD have been recently reported [42].
- The *STS* requires participants to stand up from and to sit down on a slightly padded armless chair as quickly as possible consecutively for five times. Patients fold their arms across their chests and are instructed to stand up completely while making firm contact when sitting. Timing count begins on the command "go" and ceases when the participants sit at the end of the fifth elevation up to the standing position. Subjects are allowed a practice trial of two repetitions before the recorded series of two consecutive trials of five repetitions. The faster of the two trials is then used for evaluation [43].

11.3 Rehabilitation as Process of Care

Pulmonary rehabilitation (PR) is defined as "a comprehensive intervention based on a thorough patient assessment followed by patient-tailored therapies that include, but are not limited to, exercise training, education, and behaviour change, designed to improve the physical and psychological condition of people with chronic respiratory disease and to promote the long-term adherence to health-enhancing behaviours" [44]. The two mainstays of PR are exercise training and education, followed by psychosocial support and nutritional counselling [44]. We briefly summarise in the following subparagraphs the main contents of each component in a structured programme of PR at which patients with CRD are commonly referred. Indeed, although literature developed around the impact of PR in the "COPD model", other patients suffering from respiratory disorders including asthma, cystic fibrosis and bronchiectasis, interstitial lung diseases and neuromuscular disorders involving the respiratory system are likely to potentially benefit from a rehabilitation course [45].

11.3.1 Exercise Training

Exercise training is the cornerstone of effective PR and may include several activities, such as endurance exercise training, interval exercise training, walking exercise, Nordic walking [46, 47], resistance training, aquatic exercise, classroom callisthenics and Tai Chi [48]. Although it does not change pulmonary function, exercise training improves capacity and reduces dyspnea. In order to achieve clinically relevant results, training should be strictly supervised and performed properly and for appropriate duration and frequency. Table 11.4 shows the main body sites for application of training and modalities on how to deliver exercise targeted at therapy in respiratory patients.

High-intensity exercise is more commonly employed in PR. However, patients may not be able to sustain high intensities for long time. In these cases, adherence with high-intensity training schedules may be difficult. As an alternative, low-intensity training, such as classroom callisthenics, may occasionally be considered. Furthermore, training duration and intensities vary among patients [49, 50] depending on specific deficits and individual requirements. In parallel with exercise training, improved self-efficacy resulting from education, psychosocial support and nutritional counselling (in patients with nutritional abnormalities) may lead to better long-term adherence to the training prescriptions.

Aerobic training—Aerobic training is the key component of exercise training in patients with COPD, in particular [44]. It can be performed on a cycle ergometer and/or a stationary treadmill (most frequently) but also by means of stair climbing, stepping, free walking, Nordic walking and/or swimming. In order to optimize the performance of activities of daily living [51], upper limb aerobic training can also be prescribed. High-intensity endurance-based exercise (exercise tests >10 min) is

Table 11.4 Body sites and modalities on how to deliver exercise training along PR course

Body sites	Type	Intensity	Duration of the training	Length of the programme
Lower limbs	Endurance	70–90% of the max HR or VO_2	20–45 min	3–5 times/week up to 8 weeks
	Strength	50–80% fraction of max weight lifted	8–10 rep up to 3 series	3 times/week up to 8 weeks
Upper limbs	Endurance	70–90% of the max HR or VO2	20–45 min	3–5 times/week up to 8 weeks
	Strength	50–80% fraction of max weight lifted	8–10 rep up to 3 series	3 times/week up to 8 weeks
Respiratory muscles	Strength	15–60% of MIP or MEP	1 h	3–6 times/week up to 3 months

HR heart rate, VO_2 oxygen uptake, *MIP* maximal inspiratory pressure, *MEP* maximal expiratory pressure

the main aerobic training method. Specifically, high-intensity interval training appears to be practicable even in patients with a severe respiratory disease, resulting in similar improvements of 6-min walking distance and health-related quality of life compared to traditional endurance training [52–54].

Resistance training—Resistance training is based on repetitive lifting of relatively high loads. Compared to aerobic training, resistance training produces lower cardiorespiratory responses and less dyspnea, which is highly desirable in patients with more severe CRD [55]. One to 3 sets of 812 repetitions should be performed on 2 to 3 days per week in order to reach the best results in terms of muscle strength [56]. The main results of an adequate (60–70% of one-repetition maximum [57]) high-intensity resistance training are the increased muscle mass and muscle strength, paralleled by an increased submaximal exercise tolerance [58].

Combined training and additional means of increasing exercise capacity—If on one hand aerobic training improves skeletal muscle strength and resistance training improves aerobic exercise tolerance, evidences suggest that the best results are reached by combining aerobic and resistance training together [59] and by challenging both the cardiorespiratory fitness and the muscular strength capacity.

Recent research has focused on interventions that can be used as an adjunct to exercise training in PR, especially in patients with more severe CRD and disabling breathlessness. Among these interventions, the use of supplemental oxygen and ventilator support during training was tested in COPD patients and resulted in greater improvement of exercise tolerance [60] and dyspnea [61, 62]. Probably these improvements are related to the reduction in the high inspiratory muscle load secondary to the effects of hyperinflation. Furthermore, a recent paper studied the effects of helium and oxygen (Heliox) mixtures on exercise capacity in severe COPD [63]. The rationale of using Heliox to reduce breathlessness is based on the principle that nitrogen in inspired air is substituted with helium at a lower density, which reduces resistance in the airway, improving ventilation and gas exchange.

11.3.2 Education

Education is another key component of pulmonary rehabilitation. It has gradually evolved from a didactic approach to the promotion of behaviour changes and collaborative self-management [64]. Examples of positive behaviour changes include higher adherence to medication, increased physical activity, better nutritional habits, breathing regulation techniques and applying energy-saving strategies during activities of daily living [65]. These strategies promote the self-efficacy in managing health through increasing the patients' knowledge and stimulating patients to participate with healthcare professionals in better managing their illness [66]. In Table 11.5, the main topics concerning educational component of pulmonary rehabilitation are displayed.

Table 11.5 Educational component of pulmonary rehabilitation: topics

• Normal pulmonary anatomy and physiology.
• Pathophysiology of chronic respiratory disease.
• Communicating with the healthcare provider.
• Interpretation of medical testing.
• Breathing strategies.
• Secretion clearance techniques.
• Role and rationale for medications, including oxygen therapy.
• Effective use of respiratory devices.
• Benefits of exercise and physical activities.
• Energy conservation during activities of daily living.
• Healthy food intake
• Irritant avoidance.
• Early recognition and treatment of exacerbations.
• Leisure activities.
• Coping with chronic lung disease.

11.3.3 Psychological Support

Together with education, psychological support is an integral part of PR programmes. Indeed, the incidence of depression in patients with CRD is more than twice higher compared with the general population [67]. PR programmes including psychological interventions improve the mood disorders more than those consisting of exercise training only [68]. Psychological support may be of benefit to those patients presenting with symptoms of anxiety and depression, helping them to better understand the psychological modifications that may occur in CRD [69, 70] and to encourage active participation in healthcare. Furthermore, psychologists are the best healthcare providers who can also discuss smoking cessation strategies within the course of PR and with the goal to optimise benefits.

Supervised exercise combined with stress management education and psychotherapy in PR may offer management strategies for patients with anxiety and depression [71] and may induce reduction in dyspnea sensation [72], probably due to the social interaction and distraction from negative perceptions that occur during exercise within a group of patients who have the same condition.

11.3.4 Nutritional Counselling

Nutritional counselling has a pivotal role in the PR programme for people with CRD and consists of teaching patients about how to plan and follow a healthy diet. Indeed, weight loss and body composition abnormalities are prevalent in CRD and can indirectly affect disease severity and prognosis (hospitalization and mortality) [73]. Furthermore, being undernourished in COPD is likely to be associated with longer in-patient hospital stays [74], a higher risk of being readmitted [75] and an increase in healthcare utilisation [76] in comparison with normally nourished individuals. Patients who are overweight will get advice about planning a diet that will help them to lose weight; underweight patients will receive advice about foods that can help them to gain weight. However, more cost-effectiveness studies about nutritional counselling and supplementation are still needed to support decision-making and to tackle with organisational problems, such as dealing with reimbursement for these interventions in CRD.

11.3.5 Tips and Pitfalls

Timing—Although most PR programmes enrol patients with moderate to severe CRD [77], recent studies suggest that patients with less severe degree of airflow limitation also benefit from PR programmes in terms of several outcomes. In fact, low physical activity, problems during the activities of daily living, exertional dyspnea, lower limb muscle weakness, osteoporosis, anxiety and depression may also occur in mild to moderate disease [5, 44]. Furthermore, by improving exercise tolerance and body composition and promoting self-efficacy and behaviour change, PR at an earlier stage of disease has the potential to significantly modify the course of the illness. Hence, irrespective of the degree of lung function impairment, the correct timing of PR should be rather set on the individual's clinical status and disability [44]. Therefore, early intervention and physiotherapy following clinical deterioration and/or at the very early onset of symptoms may provide substantial benefit even in these patients.

Maintenance of benefits—Without any maintenance strategy, benefits of PR tend to diminish over 6–12 months with particular regard to the physical performance. This is probably due to a decrease in adherence to regular exercise [78, 79] as well as a worsening of main disease and the clinical impact of related comorbidities [80]. Studies have examined the effects of maintenance strategies (i.e. weekly or monthly follow-up session) after PR with equivocal results about improvements in exercise tolerance/capacity and health-related quality of life [81, 82]. On the other hand, behaviour change, incorporating self-efficacy and self-management techniques, seems to be the most effective strategy for optimization and long-term maintenance of any achieved health benefit [44].

Adherence to exercise and physical activity—Monthly phone calls accompanied by a formal home programme have been shown to encourage long-term adherence to exercise, not only leading to improved walked distance and perceived health-related quality of life but also reducing lung function decline, in patients with moderate COPD after a 3-week outpatient rehabilitation [82].

Qualitative data provide further opportunities for additional peer support in patients who have completed PR, through group activities with other individuals who have similar needs and experience, including drop-in centres and exercise classes [81]. This "voluntary and mutually supportive, people like us" approach may be a valid and important alternative to regular phone calls from staff and appointments with therapists and physicians.

11.4 Conclusions

Pulmonary rehabilitation is a recognized and effective clinical process providing specific benefits to symptomatic patients with CRD, in particular those suffering from COPD. It appears essential to recognize the most appropriate programme content and setting to be delivered on an individual basis following patient's selection and referral.

To date, it is important to recognize that this therapeutic but non-pharmacological approach results in substantial effectiveness when applied at the very early onset of disability following CRD, such as during acute exacerbation of the disease [83]. Behavioural changes (i.e. improvement in long-lasting physical activity, in particular) remains a true challenge to target in the whole population of patients with CRD with the final scope to prompt interventions and limit their disability which is more and more problematic with the increasing complexity of the underlying diseases.

Notwithstanding, other perspectives are still to come in the field of PR and should be subjected to special attention from both the professionals and the stakeholders involved. Indeed, despite the evidence, there is actual low applicability, access and homogeneity of programmes across different countries [84]. Furthermore, barriers for patients should be better focused and overcome; in this light, e-health and new technologies might be helpful to achieve this goal.

Key Points

- Disability represents the hallmark of the disease important to the patient's life and must be assessed in chronic respiratory diseases (CRD).
- The mobility-related dyspnea and the resulting decrease in exercise capacity substantially contribute to increased risk of disability.

Pulmonary rehabilitation (PR) is beneficial for these symptomatic patients to limit and to counteract the progressive loss of physical function.

References

1. Harber P. Respiratory disability: what is it, how can we measure it, what causes it and is it important? Thorax. 2009;64(4):280–2.
2. Eisner MD, Iribarren C, Blanc PD, Yelin EH, Ackerson L, Byl N, et al. Development of disability in chronic obstructive pulmonary disease: beyond lung function. Thorax. 2011;66(2):108–14.
3. Barreiro E, Gea J. Respiratory and limb muscle dysfunction in COPD. COPD. 2015;12(4):413–26.
4. Maltais F, Decramer M, Casaburi R, Barreiro E, Burelle Y, Debigare R, et al. An official American Thoracic Society/European Respiratory Society statement: update on limb muscle dysfunction in chronic obstructive pulmonary disease. Am J Respir Crit Care Med. 2014;189(9):e15–62.

5. Seymour JM, Spruit MA, Hopkinson NS, Natanek SA, Man WD, Jackson A, et al. The prevalence of quadriceps weakness in COPD and the relationship with disease severity. Eur Respir J. 2010;36(1):81–8.
6. Shrikrishna D, Patel M, Tanner RJ, Seymour JM, Connolly BA, Puthucheary ZA, et al. Quadriceps wasting and physical inactivity in patients with COPD. Eur Respir J. 2012;40(5):1115–22.
7. Bernard S, LeBlanc P, Whittom F, Carrier G, Jobin J, Belleau R, et al. Peripheral muscle weakness in patients with chronic obstructive pulmonary disease. Am J Respir Crit Care Med. 1998;158(2):629–34.
8. Gosselink R, Troosters T, Decramer M. Distribution of muscle weakness in patients with stable chronic obstructive pulmonary disease. J Cardiopulm Rehabil. 2000;20(6):353–60.
9. Hopkinson NS, Tennant RC, Dayer MJ, Swallow EB, Hansel TT, Moxham J, et al. A prospective study of decline in fat free mass and skeletal muscle strength in chronic obstructive pulmonary disease. Respir Res. 2007;8(1):25.
10. Spruit MA, Gosselink R, Troosters T, Kasran A, Gayan-Ramirez G, Bogaerts P, et al. Muscle force during an acute exacerbation in hospitalised patients with COPD and its relationship with CXCL8 and IGF-I. Thorax. 2003;58(9):752–6.
11. Osthoff AK, Taeymans J, Kool J, Marcar V, van Gestel AJ. Association between peripheral muscle strength and daily physical activity in patients with COPD: a systematic literature review and meta-analysis. J Cardiopulm Rehabil Prev. 2013;33(6):351–9.
12. Butcher SJ, Pikaluk BJ, Chura RL, Walkner MJ, Farthing JP, Marciniuk DD. Associations between isokinetic muscle strength, high-level functional performance, and physiological parameters in patients with chronic obstructive pulmonary disease. Int J Chron Obstruct Pulmon Dis. 2012;7:537–42.
13. Patel MS, Mohan D, Andersson YM, Baz M, Samantha Kon SC, Canavan JL, et al. Phenotypic characteristics associated with reduced short physical performance battery score in COPD. Chest. 2014;145(5):1016–24.
14. Gosselink R, Troosters T, Decramer M. Peripheral muscle weakness contributes to exercise limitation in COPD. Am J Respir Crit Care Med. 1996;153(3):976–80.
15. Swallow EB, Reyes D, Hopkinson NS, Man WD, Porcher R, Cetti EJ, et al. Quadriceps strength predicts mortality in patients with moderate to severe chronic obstructive pulmonary disease. Thorax. 2007;62(2):115–20.
16. Nyberg A, Saey D, Maltais F. Why and how limb muscle mass and function should be measured in patients with chronic obstructive pulmonary disease. Ann Am Thorac Soc. 2015;12(9):1269–77.
17. Engelen MP, Schols AM, Does JD, Wouters EF. Skeletal muscle weakness is associated with wasting of extremity fat-free mass but not with airflow obstruction in patients with chronic obstructive pulmonary disease. Am J Clin Nutr. 2000;71(3):733–8.
18. Schols AM, Soeters PB, Dingemans AM, Mostert R, Frantzen PJ, Wouters EF. Prevalence and characteristics of nutritional depletion in patients with stable COPD eligible for pulmonary rehabilitation. Am Rev Respir Dis. 1993;147(5):1151–6.
19. Baarends EM, Schols AM, Mostert R, Wouters EF. Peak exercise response in relation to tissue depletion in patients with chronic obstructive pulmonary disease. Eur Respir J. 1997;10(12):2807–13.
20. Kobayashi A, Yoneda T, Yoshikawa M, Ikuno M, Takenaka H, Fukuoka A, et al. The relation of fat-free mass to maximum exercise performance in patients with chronic obstructive pulmonary disease. Lung. 2000;178(2):119–27.
21. Schols AM, Mostert R, Soeters PB, Wouters EF. Body composition and exercise performance in patients with chronic obstructive pulmonary disease. Thorax. 1991;46(10):695–9.
22. Mostert R, Goris A, Weling-Scheepers C, Wouters EF, Schols AM. Tissue depletion and health related quality of life in patients with chronic obstructive pulmonary disease. Respir Med. 2000;94(9):859–67.
23. Marquis K, Debigare R, Lacasse Y, LeBlanc P, Jobin J, Carrier G, et al. Midthigh muscle cross-sectional area is a better predictor of mortality than body mass index in patients with chronic obstructive pulmonary disease. Am J Respir Crit Care Med. 2002;166(6):809–13.

24. de Blasio F, Santaniello MG, de Blasio F, Mazzarella G, Bianco A, Lionetti L, et al. Raw BIA variables are predictors of muscle strength in patients with chronic obstructive pulmonary disease. Eur J Clin Nutr. 2017;71(11):1336–40.
25. Maddocks M, Kon SS, Jones SE, Canavan JL, Nolan CM, Higginson IJ, et al. Bioelectrical impedance phase angle relates to function, disease severity and prognosis in stable chronic obstructive pulmonary disease. Clin Nutr. 2015;34(6):1245–50.
26. de Blasio F, Scalfi L, Alicante P, Miracco Berlingieri G, Bellofiore B, de Blasio F. Malnutrition and sarcopenia in chronic obstructive pulmonary disease according to the new ESPEN definition and EWGSOP criteria. Eur Respir J. 2017;50(Suppl 61):PA400.
27. de Blasio F, Scalfi L, Di Gregorio A, Miracco Berlingieri G, De Martino M, de Blasio F. Raw bioelectrical impedance variables are predictors of survival in chronic obstructive pulmonary disease. Eur Respir J. 2017;50(Suppl 61):PA1094.
28. St-Onge MP, Wang J, Shen W, Wang Z, Allison DB, Heshka S, et al. Dual-energy x-ray absorptiometry-measured lean soft tissue mass: differing relation to body cell mass across the adult life span. J Gerontol A Biol Sci Med Sci. 2004;59(8):796–800.
29. Bestall J, Paul E, Garrod R, Garnham R, Jones P, Wedzicha J. Usefulness of the Medical Research Council (MRC) dyspnoea scale as a measure of disability in patients with chronic obstructive pulmonary disease. Thorax. 1999;54(7):581–6.
30. Sundh J, Janson C, Lisspers K, Ställberg B, Montgomery S. Dyspnoea, obstruction, smoking and exacerbation (DOSE) index and mortality in COPD. Eur Respir J. 2011;38(Suppl 55):542S.
31. Borg G. Ratings of perceived exertion and heart rates during short-term cycle exercise and their use in a new cycling strength test. Int J Sports Med. 1982;3(3):153–8.
32. Aitken RC. Measurement of feelings using visual analogue scales. Proc R Soc Med. 1969;62(10):989–93.
33. Goldstein RE. In: Walker HK, Hall WD, Hurst JW, editors. Clinical methods: The History P, and Laboratory Examinations. 3rd ed. Boston: Butterworths; 1990. Chapter 8.
34. Palange P, Ward SA, Carlsen KH, Casaburi R, Gallagher CG, Gosselink R, et al. Recommendations on the use of exercise testing in clinical practice. Eur Respir J. 2007;29(1):185–209.
35. Weisman IM, Zeballos RJ. An integrated approach to the interpretation of cardiopulmonary exercise testing. Clin Chest Med. 1994;15(2):421–45.
36. Sue DY, Wasserman K. Impact of integrative cardiopulmonary exercise testing on clinical decision making. Chest. 1991;99(4):981–92.
37. Ferguson B. ACSM's guidelines for exercise testing and prescription 9th Ed. 2014. J Can Chiropr Assoc. 2014;58(3):328.
38. ATS Committee on Proficiency Standards for Clinical Pulmonary Function Laboratories. ATS statement: guidelines for the six-minute walk test. Am J Respir Crit Care Med. 2002;166(1):111–7.
39. Borg GA. Psychophysical bases of perceived exertion. Med Sci Sports Exerc. 1982;14(5):377–81.
40. Singh SJ, Morgan MD, Scott S, Walters D, Hardman AE. Development of a shuttle walking test of disability in patients with chronic airways obstruction. Thorax. 1992;47(12):1019–24.
41. Revill S, Morgan M, Singh S, Williams J, Hardman A. The endurance shuttle walk: a new field test for the assessment of endurance capacity in chronic obstructive pulmonary disease. Thorax. 1999;54(3):213–22.
42. Albarrati AM, Gale NS, Enright S, Munnery MM, Cockcroft JR, Shale DJ. A simple and rapid test of physical performance in chronic obstructive pulmonary disease. Int J Chron Obstruct Pulmon Dis. 2016;11:1785–91.
43. Bohannon RW, Bubela DJ, Magasi SR, Wang YC, Gershon RC. Sit-to-stand test: performance and determinants across the age-span. Isokinet Exerc Sci. 2010;18(4):235–40.
44. Spruit MA, Singh SJ, Garvey C, ZuWallack R, Nici L, Rochester C, et al. An official American Thoracic Society/European Respiratory Society statement: key concepts and advances in pulmonary rehabilitation. Am J Respir Crit Care Med. 2013;188(8):e13–64.

45. Holland AE, Wadell K, Spruit MA. How to adapt the pulmonary rehabilitation programme to patients with chronic respiratory disease other than COPD. Eur Respir Rev. 2013;22:577–86.
46. Breyer MK, Breyer-Kohansal R, Funk GC, Dornhofer N, Spruit MA, Wouters EF, et al. Nordic walking improves daily physical activities in COPD: a randomised controlled trial. Respir Res. 2010;11:112.
47. Dolmage TE, Goldstein RS. Effects of one-legged exercise training of patients with COPD. Chest. 2008;133(2):370–6.
48. Andrianopoulos V, Klijn P, Franssen FM, Spruit MA. Exercise training in pulmonary rehabilitation. Clin Chest Med. 2014;35(2):313–22.
49. Normandin EA, McCusker C, Connors M, Vale F, Gerardi D, ZuWallack RL. An evaluation of two approaches to exercise conditioning in pulmonary rehabilitation. Chest. 2002;121(4):1085–91.
50. Casaburi R, Patessio A, Ioli F, Zanaboni S, Donner CF, Wasserman K. Reductions in exercise lactic acidosis and ventilation as a result of exercise training in patients with obstructive lung disease. Am Rev Respir Dis. 1991;143(1):9–18.
51. Janaudis-Ferreira T, Hill K, Goldstein R, Wadell K, Brooks D. Arm exercise training in patients with chronic obstructive pulmonary disease: a systematic review. J Cardiopulm Rehabil Prev. 2009;29(5):277–83.
52. Vogiatzis I, Nanas S, Roussos C. Interval training as an alternative modality to continuous exercise in patients with COPD. Eur Respir J. 2002;20(1):12–9.
53. Varga J, Porszasz J, Boda K, Casaburi R, Somfay A. Supervised high intensity continuous and interval training vs. self-paced training in COPD. Respir Med. 2007;101(11):2297–304.
54. Kortianou EA, Nasis IG, Spetsioti ST, Daskalakis AM, Vogiatzis I. Effectiveness of interval exercise training in patients with COPD. Cardiopulm Phys Ther J. 2010;21(3):12–9.
55. Probst VS, Troosters T, Pitta F, Decramer M, Gosselink R. Cardiopulmonary stress during exercise training in patients with COPD. Eur Respir J. 2006;27(6):1110–8.
56. Garber CE, Blissmer B, Deschenes MR, Franklin BA, Lamonte MJ, Lee IM, et al. American College of Sports Medicine position stand. Quantity and quality of exercise for developing and maintaining cardiorespiratory, musculoskeletal, and neuromotor fitness in apparently healthy adults: guidance for prescribing exercise. Med Sci Sports Exerc. 2011;43(7):1334–59.
57. Gloeckl R, Marinov B, Pitta F. Practical recommendations for exercise training in patients with COPD. Eur Respir Rev. 2013;22(128):178–86.
58. O'Shea SD, Taylor NF, Paratz J. Peripheral muscle strength training in COPD: a systematic review. Chest. 2004;126(3):903–14.
59. Ortega F, Toral J, Cejudo P, Villagomez R, Sanchez H, Castillo J, et al. Comparison of effects of strength and endurance training in patients with chronic obstructive pulmonary disease. Am J Respir Crit Care Med. 2002;166(5):669–74.
60. Emtner M, Porszasz J, Burns M, Somfay A, Casaburi R. Benefits of supplemental oxygen in exercise training in nonhypoxemic chronic obstructive pulmonary disease patients. Am J Respir Crit Care Med. 2003;168(9):1034–42.
61. Hawkins P, Johnson LC, Nikoletou D, Hamnegard CH, Sherwood R, Polkey MI, et al. Proportional assist ventilation as an aid to exercise training in severe chronic obstructive pulmonary disease. Thorax. 2002;57(10):853–9.
62. Garrod R, Mikelsons C, Paul EA, Wedzicha JA. Randomized controlled trial of domiciliary noninvasive positive pressure ventilation and physical training in severe chronic obstructive pulmonary disease. Am J Respir Crit Care Med. 2000;162(4 Pt 1):1335–41.
63. Wedzicha JA. Heliox in chronic obstructive pulmonary disease. Am J Respir Crit Care Med. 2006;173(8):825–6.
64. Mazzuca SA. Does patient education in chronic disease have therapeutic value? J Chronic Dis. 1982;35(7):521–9.
65. Velloso M, Jardim JR. Study of energy expenditure during activities of daily living using and not using body position recommended by energy conservation techniques in patients with COPD. Chest. 2006;130(1):126–32.
66. Bourbeau J, Nault D, Dang-Tan T. Self-management and behaviour modification in COPD. Patient Educ Couns. 2004;52(3):271–7.

67. van Manen JG, Bindels PJ, Dekker FW, IJzermans C, van der Zee JS, Schade E. Risk of depression in patients with chronic obstructive pulmonary disease and its determinants. Thorax. 2002;57(5):412–6.
68. de Godoy DV, de Godoy RF. A randomized controlled trial of the effect of psychotherapy on anxiety and depression in chronic obstructive pulmonary disease. Arch Phys Med Rehabil. 2003;84(8):1154–7.
69. Casaburi R, ZuWallack R. Pulmonary rehabilitation for management of chronic obstructive pulmonary disease. N Engl J Med. 2009;360(13):1329–35.
70. Norwood R. Prevalence and impact of depression in chronic obstructive pulmonary disease patients. Curr Opin Pulm Med. 2006;12(2):113–7.
71. Harrison SL, Greening NJ, Williams JE, Morgan MD, Steiner MC, Singh SJ. Have we underestimated the efficacy of pulmonary rehabilitation in improving mood? Respir Med. 2012;106(6):838–44.
72. Haas F, Salazar-Schicchi J, Axen K. Desensitization to dyspnea in chronic obstructive pulmonary disease. In: Casaburi R, Petty TL, editors. Principles and practice of pulmonary rehabilitation. Philadelphia: W.B. Saunders; 1993. p. 241–51.
73. Vestbo J, Prescott E, Almdal T, Dahl M, Nordestgaard BG, Andersen T, et al. Body mass, fat-free body mass, and prognosis in patients with chronic obstructive pulmonary disease from a random population sample: findings from the Copenhagen City Heart Study. Am J Respir Crit Care Med. 2006;173(1):79–83.
74. Gupta B, Kant S, Mishra R, Verma S. Nutritional status of chronic obstructive pulmonary disease patients admitted in hospital with acute exacerbation. J Clin Med Res. 2010;2(2): 68–74.
75. Hallin R, Koivisto-Hursti UK, Lindberg E, Janson C. Nutritional status, dietary energy intake and the risk of exacerbations in patients with chronic obstructive pulmonary disease (COPD). Respir Med. 2006;100(3):561–7.
76. Odencrants S, Ehnfors M, Ehrenberg A. Nutritional status and patient characteristics for hospitalised older patients with chronic obstructive pulmonary disease. J Clin Nurs. 2008;17(13):1771–8.
77. Lacasse Y, Goldstein R, Lasserson TJ, Martin S. Pulmonary rehabilitation for chronic obstructive pulmonary disease. Cochrane Database Syst Rev. 2006;4:Cd003793.
78. Griffiths TL, Burr ML, Campbell IA, Lewis-Jenkins V, Mullins J, Shiels K, et al. Results at 1 year of outpatient multidisciplinary pulmonary rehabilitation: a randomised controlled trial. Lancet. 2000;355(9201):362–8.
79. Spruit MA, Troosters T, Trappenburg JC, Decramer M, Gosselink R. Exercise training during rehabilitation of patients with COPD: a current perspective. Patient Educ Couns. 2004;52(3):243–8.
80. Carr SJ, Goldstein RS, Brooks D. Acute exacerbations of COPD in subjects completing pulmonary rehabilitation. Chest. 2007;132(1):127–34.
81. Steele BG, Belza B, Cain KC, Coppersmith J, Lakshminarayan S, Howard J, et al. A randomized clinical trial of an activity and exercise adherence intervention in chronic pulmonary disease. Arch Phys Med Rehabil. 2008;89(3):404–12.
82. Waterhouse JC, Walters SJ, Oluboyede Y, Lawson RA. A randomised 2 × 2 trial of community versus hospital pulmonary rehabilitation, followed by telephone or conventional follow-up. Health Technol Assess. 2010;14(6.):i–v, vii–xi):1–140.
83. Spruit MA, Singh SJ, Rochester CL, Greening NJ, Franssen FME, Pitta F, et al. Pulmonary rehabilitation for patients with COPD during and after an exacerbation-related hospitalisation: back to the future? Eur Respir J. 2018;51:1.
84. Vogiatzis I, Rochester CL, Spruit MA, Troosters T, Clini EM. Increasing implementation and delivery of pulmonary rehabilitation: key messages from the new ATS/ERS policy statement. Eur Respir J. 2016;47(5):1336–41.
85. Robles PG, Mathur S, Janaudis-Fereira T, Dolmage TE, Goldstein RS, Brooks D. Measurement of peripheral muscle strength in individuals with chronic obstructive pulmonary disease: a systematic review. J Cardiopulm Rehabil Prev. 2011;31(1):11–24.

86. Bachasson D, Wuyam B, Pepin JL, Tamisier R, Levy P, Verges S. Quadriceps and respiratory muscle fatigue following high-intensity cycling in COPD patients. PLoS One. 2013;8(12):e83432.
87. Evans RA, Kaplovitch E, Beauchamp MK, Dolmage TE, Goldstein RS, Gillies CL, et al. Is quadriceps endurance reduced in COPD? Chest. 2015;147(3):673–84.
88. Vieira L, Bottaro M, Celes R, Viegas CA, Silva C. Isokinetic muscle evaluation of quadriceps in patients with chronic obstructive pulmonary disease. Rev Port Pneumol. 2010;16(5):717–36.
89. Thompson WR, Gordon NF, Pescatello LS. ACSM's guidelines for exercise testing and prescription. Philadelphia: Wolters Kluwer/Lippincott Williams & Wilkins; 2010.
90. Couillard A, Koechlin C, Cristol JP, Varray A, Prefaut C. Evidence of local exercise-induced systemic oxidative stress in chronic obstructive pulmonary disease patients. Eur Respir J. 2002;20(5):1123–9.
91. Couillard A, Maltais F, Saey D, Debigaré R, Michaud A, Koechlin C, et al. Exercise-induced quadriceps oxidative stress and peripheral muscle dysfunction in patients with COPD. Am J Respir Crit Care Med. 2003;167:1664–9.
92. Stark T, Walker B, Phillips JK, Fejer R, Beck R. Hand-held dynamometry correlation with the gold standard isokinetic dynamometry: a systematic review. PM R. 2011;3(5):472–9.
93. O'Shea SD, Taylor NF, Paratz JD. Measuring muscle strength for people with chronic obstructive pulmonary disease: retest reliability of hand-held dynamometry. ArchPhysMed Rehabil. 2007;88(1):32–6.
94. Hartmann A, Knols R, Murer K, De Bruin ED. Reproducibility of an isokinetic strength-testing protocol of the knee and ankle in older adults. Gerontology. 2009;55(3):259–68.
95. Dourado VZ. Relationship of upper-limb and thoracic muscle strength to 6-min walk distance in COPD patients. Chest. 2006;129(3):551–7.
96. Burns SP, Spanier DE. Break-technique handheld dynamometry: relation between angular velocity and strength measurements. ArchPhysMed Rehabil. 2005;86(7):1420–6.
97. Burns SP, Breuninger A, Kaplan C, Marin H. Hand-held dynamometry in persons with tetraplegia: comparison of make- versus break-testing techniques. Am J Phys Med Rehabil. 2005;84(1):22–9.
98. Bachasson D, Villiot-Danger E, Verges S, Hayot M, Perez T, Chambellan A, et al. Maximal isometric voluntary quadriceps strength assessment in COPD. Revue des Maladies Respiratoires. 2014;31(8):765–70.

Part V
Special Considerations

Lung and Exercise in Extreme Environments

12

Annalisa Cogo, Maurizio Schiavon, and Lorenza Pratali

Abstract

Changes in barometric pressure characterize both high altitude and the underwater world. These changes have a significant impact on the body's pulmonary physiology and can affect both mountain climbers and divers. In both of these conditions, the organism must use compensatory mechanisms to adapt to the different environmental conditions. In this chapter, the environmental characteristics of altitude and underwater and their effects on the respiratory system are discussed. The body's compensatory responses at high altitude, primarily the ventilatory response, are discussed. A section is devoted to the maladaptive response (i.e. high-altitude pulmonary oedema) and its prevention and treatment. Next, we consider the safety of high-altitude travel in patients with chronic respiratory disease, primarily chronic obstructive pulmonary disease, asthma and cystic fibrosis. In the second part of the chapter, the physiology and pathophysiology of the hyperbaric and diving environments, especially those involving the respiratory system, are described. A paragraph is dedicated to fitness for immersion, especially for the most common chronic respiratory diseases, such as asthma and COPD. In the last part of the chapter, we discuss lung interstitial oedema, its development in extreme environments or after strenuous exercise, and the diagnostic methods.

A. Cogo (✉)
Center for Exercise and Sport Science, University of Ferrara, Ferrara, Italy
e-mail: annalisa.cogo@unife.it

M. Schiavon
Physical Activity Unit—Sports Medicine, Padova, Italy
e-mail: maurizio.schiavon@aulss6.veneto.it

L. Pratali
Institute of Clinical Physiology, CNR, Pisa, Italy
e-mail: lorenza@ifc.cnr.it

A. Cogo et al. (eds.), *Exercise and Sports Pulmonology*,
https://doi.org/10.1007/978-3-030-05258-4_12

12.1 Lung and High Altitude

12.1.1 Introduction

High altitude is characterized by a progressive reduction of barometric pressure with a consequent progressive reduction of inspiratory oxygen pressure. Even if this reduced barometric pressure is the key factor of altitude, other environmental changes can affect the lung: the progressive reduction of the air density, of the absolute humidity and of the temperature. Ultimately, at altitude the human body is exposed to progressive hypoxia and breathes a progressively drier and colder air. Based on the effects of altitude on human physiology and well-being in healthy individuals, altitude is generally divided into low altitude, moderate/intermediate altitude, high altitude and extreme altitude [1, 2]. *Extreme altitude* is the highest altitude at which long-term human adaptation is possible, as the highest permanent settlements are at this altitude [3]. The physiological changes due to hypoxia, the reduction of exercise capacity, the possible development of altitude illness and the need for acclimatization increase with increasing altitude.

12.1.2 Hypoxic Profile at Altitude

The continuous 24-h monitoring of oxygen saturation during the ascent to 5050 m has shown that the time spent with oxygen saturation $\leq 90\%$ increases with altitude. In particular, at 3500 m, almost 75% of the time is spent with $SpO_2 < 90\%$, while at 5000 m it is 94% [4]. The prediction of the "normal" reduction of oxygen saturation at increasing altitude is difficult: from 2000 to 6000 m, an average decline of 5.75%/1000 m is expected [5]. Lorente-Aznar et al. have recently published a formula to predict oxygen saturation in healthy subjects at altitude: $SaO_2 = 103.3 - (\text{altitude} \times 0.0047) + 0.7$ (for males) or + 1.4 (for females) [6].

12.1.3 Compensatory Responses

Going to altitude, individuals need to adapt to the progressive hypoxia; this process is known as acclimatization and involves a series of adjustments occurring over a period of minutes, hours or months. All these compensatory responses aim at increasing the efficiency of oxygen uptake, oxygen transport and oxygen delivery to cells. Immediate adjustments involve the respiratory and cardiovascular systems and the oxygen-carrying capacity of the blood. The threshold altitude and the magnitude of the responses have a large interindividual variability [1]. The lung is the interface between the environment and the metabolic function of the body and plays a pivotal role in the acclimatization process. However, it may also be involved in maladaptive responses. The physiologic responses are the ventilatory response and hypoxic pulmonary vasoconstriction (HPV), which causes an increase in pulmonary vascular resistance and a consequent increase in pulmonary artery pressure. The maladaptive response is the development of high-altitude pulmonary oedema (HAPE).

12.1.3.1 Hypoxic Ventilatory Response (HVR)

The respiratory compensatory response is an increase in ventilation, which means an increase in both tidal volume and breathing rate: the respiration is deeper and faster and results in higher alveolar and arterial PO_2 and lower PCO_2 levels than would be obtained if the ventilation were unchanged. HVR is due to a sensing system of the peripheral chemoreceptors, mainly the carotid body. The HVR is not a simple linear response, as it is complicated by the effect of hyperventilation on PCO_2 [7]. The extremely high work of the lung at very high altitude is well summarized by the vivid description of the difficulties exercising at extreme altitude, written by two mountaineers, climbing above 8000 m without supplementary oxygen. Norton in 1925 wrote: "Our pace was wretched. My ambition was to do 20 consecutive paces uphill mountain without a pause to rest and pant elbow on bent knee, yet I never remember achieving it- 13 was nearer the mark" [8], and Messner in 1979 wrote: "I am nothing more than a single narrow gasping lung, floating over the mists and summits" [9]. A very high ventilatory response to hypoxia is important to reach extreme altitude without supplementary oxygen, but the excessive ventilation could be counterproductive by exhausting the ventilatory reserve early on, at least at extreme altitude. Actually, studies performed on elite mountaineers climbing above 8000 m without supplementary oxygen have shown that those who have reached the summit without oxygen had a brisk but not excessive ventilatory stimulus, so they maintained a sustainable ventilation and a sufficient ventilatory reserve to climb up to extreme altitudes without oxygen supplementation [10, 11].

12.1.3.2 Breathing Pattern

Moreover, elite climbers spontaneously adopted a slower and deeper breathing pattern during exercise; this breathing strategy allowed them to maintain a higher ventilatory efficiency (i.e. the amount of ventilation required to achieve a given level of oxygen saturation: SaO_2/VE). In addition to the reported studies on elite climbers, other authors highlighted the importance of the breathing pattern during altitude exposure. Bernardi et al. compared the breathing pattern of Tibetan monks, Caucasians trained in yoga breathing and control subjects without expertise in yoga. During altitude exposure, both monks and the yoga-trained adopted a deeper and slower breathing pattern, allowing them to maintain a better oxygen saturation and a lower VE/SaO_2 ratio [12]. Bilo et al. showed that above 4000 m, the change in breathing pattern from a spontaneous rate to a paced frequency of 6 breaths per minute was associated with a significant increase in blood oxygen saturation [13].

12.1.3.3 Thoraco-Abdominal Coordination

Another important point to take into account during exercise, especially at altitude, is the thoraco-abdominal coordination. In fact, during exercise in the mountain, the postural tasks related to the presence of steep slopes and different types of terrain may modify the coordinated action of the diaphragm and the abdominal muscles.

A reduction in thoraco-abdominal coordination may in turn alter ventilatory patterns and reduce ventilatory efficiency. It has been recently reported that during intense exercise between 2000 and 2800 m, steep slopes (above 20–30%) led to a reduction in thoraco-abdominal coordination, resulting in a less efficient ventilatory pattern, decreased SpO_2 and reduced running speed [14].

12.1.4 Pulmonary Mechanics

The main changes in pulmonary mechanics described at high altitude are the increase in flows and the reduction in airway resistances, consistent with the reduced density of the air, and a variable decrease in vital capacity. The decrease in vital capacity can be due to many factors, including a reduction of respiratory muscle strength, pulmonary vascular engorgement and interstitial oedema [15–18]. However, the decrease in vital capacity and in the flows in the last part of expiration is not a hallmark of small airway compression; therefore, the attention has been focused on the functional tests, which can evaluate the small airways. The two recommended tests to evaluate distal lung function are the nitrogen single-breath test, which allows measuring the closing volume, and the measurement of respiratory impedance by the high-frequency oscillation technique [19]. Closing volume was significantly increased in 75% of almost 250 healthy climbers after arrival at 4559 m [20]. Similar results were obtained in the same setting after 24 and 48 h of stay, and the authors concluded that these alterations of pulmonary function are consistent with interstitial fluid accumulation but do not indicate a subsequent progression to clinical HAPE [21]. Regarding the oscillation technique, Pellegrino et al. reported a significant change in reactance in healthy subjects during exposure to 4559 m [22], confirming data previously obtained in experimental animals developing interstitial oedema [23]. Reactance is an index reflecting the elastic properties of the tissues or the resistance of distal airways [24]. This reduction in small airway function can be interpreted as due to the development of interstitial lung oedema (or subclinical lung oedema) [25]. A paragraph will be dedicated to this topic in the last part of the chapter.

12.1.5 High-Altitude Pulmonary Oedema (HAPE)

HAPE is a non-cardiogenic lung oedema due to the combination of different factors; the pathophysiology is complex, and an extensive discussion can be found elsewhere [26, 27]. The hallmark is undoubtedly an excessive HPV. In fact, when exposed to hypoxia, HAPE-susceptible individuals have an exaggerated pulmonary vasoconstriction, resulting in higher pulmonary pressure both at rest and during exercise. Furthermore, an impaired alveolar reabsorption and epithelial ion transport capacity and increased endothelial permeability contribute to the development of the disease. As there is a large interindividual variability in HPV, it follows that individual susceptibility is the most important determinant of the occurrence of HAPE. Other risk factors are a too fast ascent profile and overexertion just after arriving at high altitude. HAPE affects 0.2–8% of travellers between 2500 and 5500 m, with greater incidence at higher altitude, especially if reached with a fast ascent. Onset is usually within 2–5 days of permanence at high altitude. Symptoms associated with HAPE are excessive fatigue for a given effort, dyspnoea at the minimal effort, orthopnoea, chest tightness and dry cough. Immediate descent (at least 1000 m or until symptoms resolve) is the most effective treatment option. If descent is not immediately possible, supplemental oxygen at a flow rate that raises the oxygen saturation to ≥90% and nifedipine administration (30 mg oral extended release twice daily) are recommended [28, 29].

The major preventive measure is a slow ascent, such as an increase in altitude not exceeding 300 m/day above an altitude of 2500 m. This rule is effective even in susceptible individuals [26]. If progressive high-altitude acclimatization is not possible, prophylaxis with nifedipine should be recommended.

12.1.6 Exposure to High Altitude for Patients Suffering from Chronic Respiratory Diseases

When we have to advise patients about altitude exposure, the main question is: does the mountain environment (hypoxia plus other features) adversely affect the disease? Furthermore, is the patient able to adapt to a high-altitude environment [30]? The literature about lowlanders suffering from chronic respiratory disease during altitude exposure is very limited (especially for COPD patients), and insight into this issue mainly comes from laboratory studies (i.e. simulated altitude) or studies with a small number of subjects (Fig. 12.1).

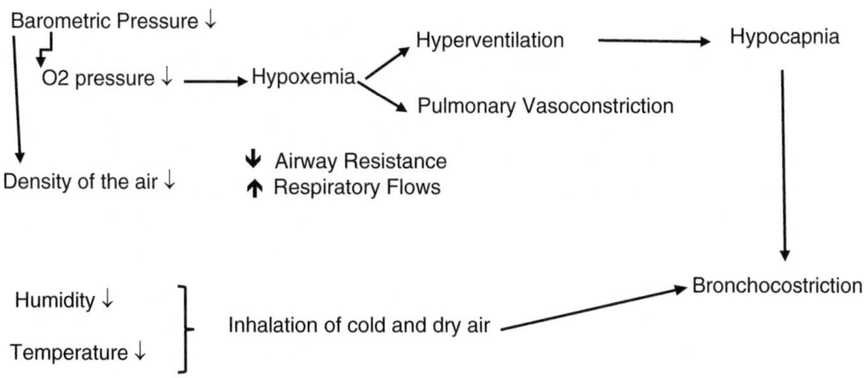

Fig. 12.1 The lung and the mountain environment. In this figure, the effects on the respiratory system of the changes of different atmospheric variables are represented

12.1.6.1 Asthma

Several features of high-altitude environments can affect asthma control, both positively and negatively. At altitude, hyperventilation of cold air, especially during exercise, has a considerable potential for airway dehydration, which can trigger an asthma attack [31]. Regarding other risk factors, pollen exposure depends on season, altitude and latitude, while dust mite exposure can be reduced due to the lower humidity. In the last 20 years, many studies have shown that a short stay at low-moderate altitude (1600–2000 m) improves clinical and functional parameters and decreases corticosteroid requirement in asthmatics [32]. Regarding higher altitudes, very few studies have assessed bronchial hyperresponsiveness in asthmatics at high altitude, but they have shown a reduction in bronchial response to both methacholine and hypoosmolar aerosol above 3500 m and up to 5000 m compared to sea level. Only a small group of mild, stable asthmatics was studied, and caution is necessary in generalizing these results [33]. It should, however, be emphasized that observational studies did not report asthma worsening during climbs and treks, at least up to an altitude of approximately 6000 m [34, 35]. Similar results were obtained among the workers engaged in the construction of the railway between Beijing and Lhasa at altitudes above 4000 m: asthmatic subjects generally did better at altitude compared to low altitude [36]. The two independent risk factors for asthma attacks during travel at altitude are the frequent use (>3 times weekly) of inhaled bronchodilators before travel and the participation in intense physical exertion during treks [37]. In conclusion, patients with asthma can travel to high altitude when their asthma is well controlled. If not, therapy must be optimized before leaving. Patients should continue regular therapy, should always have rescue drugs, and should use premedication prior to exercise with the same drugs used at sea level. As at sea level, during very cold and windy days, patients should protect the mouth (e.g. with a scarf) to warm and humidify inhaled air. Asthmatic patients with severe disease or those experiencing or recovering from an asthma exacerbation should avoid travelling to high altitude, particularly into remote areas where health facilities and medical care can be difficult to reach [30, 38].

12.1.6.2 Chronic Obstructive Pulmonary Disease (COPD)

COPD is characterized by many pathophysiological problems, which can be affected by altitude. The main questions are: Is the COPD patient able to maintain an adequate PaO_2, or does he need supplemental oxygen? Can the COPD increase the work of ventilation during hypoxic exposure?

More than 20 years ago, Dillard [39] published a formula to predict the level of hypoxemia at an altitude of 2300–2500 m (the maximum altitude allowed for an aircraft cabin), combining sea-level FEV_1 values with the sea-level PaO_2: PaO_2, Alt $= (0.519 \times Pa,O_2,SL) + (11.85 \times FEV_1) - 1.76$.

Based on subsequent research, it is now recommended to follow the flow chart published by the British Thoracic Society as a preflight evaluation [40]. The studies mentioned above were performed during altitude simulation tests, i.e. during inhalation of an hypoxic mixture simulating 2300 m. Very few studies have been performed at real altitude, totalling a small number of subjects. These studies reported at 2000 m a reduction of mean PaO_2 (from 75 ± 9 mmHg to 51 ± 6) and a reduction of the distance covered during a 6-minute walking test by 52% [41]. At higher altitude (2950 m), the endurance time decreased almost 50% (from 500 to 205 s), with severe oxygen desaturation and reduced cerebral oxygenation [42].

To summarize: How to assess a COPD patient before an altitude ascent?

- Spirometry to assess the degree of bronchial obstruction
- Blood gas analysis to assess the efficiency of gas exchange
- Six-minute walking distance to detect the presence of oxygen desaturation during exercise at sea level
- If available, hypoxia altitude simulation test (15% O_2) to predict SpO_2 at altitude

Some specific data should be gathered:

- The altitude of destination and the sleeping altitude
- The rate of ascent (by cable car, bus, walking)
- The duration of stay
- The amount of exercise at high altitude

The results of the tests and the combination of information will allow practitioners to provide advice tailored on the patient's pathophysiological characteristics.

12.1.6.3 Cystic Fibrosis

The role of exercise in cystic fibrosis (CF) is well established, and over the last three decades it has become an important component in the management of all individuals with CF [43]. Regarding exercise at altitude, tests at simulated altitude have established that CF patients with baseline $PaO_2 > 8.0$ kPa can safely tolerate an altitude of approximately 2500 m, but only under resting conditions [44]. In fact, during even mild exercise (30 watts), oxygen saturation decreases dramatically. Two studies at 1500 m have shown that a 4-week exercise training improved exercise tolerance and quality of life of CF patients, but they were at significant risk for oxygen desaturation during exercise. The lower the respiratory function (mainly the FEV_1 as a percentage of predicted), the lower the oxygen saturation at peak exercise. Therefore, CF patients must be studied carefully with a supervised exercise test before engaging in exercise at altitude and during exercise. The oxygen saturation should be monitored and maintained ≥88% [45, 46].

12.2 Lungs and Diving

12.2.1 The Physiology and Pathophysiology of the Hyperbaric and Diving Environments

12.2.1.1 Introduction

The techniques used by divers date back to antiquity, when primitive men dove underwater using their capacity to hold their breath, devoting themselves to the search for food (underwater fishing) and wealth (sponges and apnoea pearl fishing). Diving has also served military purposes: Thucydides reports how during the Athenian attack in Syracuse in 414 BC the intervention of some divers was fundamental to cut the immersed barriers that had been built to protect the harbour. The fascination with free diving remains intact even today, thanks to the feeling of freedom offered by the very limited equipment needed (mask, fins and snorkel). At the same time, it gives anyone the possibility to approach the underwater world. The inability to get oxygen underwater imposed by human physiology was also overcome thanks to the self-contained underwater breathing apparatus (SCUBA), containing air or gas mixtures with various oxygen and nitrogen concentrations. SCUBA, an open-circuit breathing apparatus, was made by Cousteau and Gagnan serving in the French Navy in 1943. Diving (SCUBA diving) can therefore be carried out with "normal" breathing. It must be limited in time, depending on the gas supply, never forgetting the constraint of physical and physiological modifications of the hyperbaric underwater environment. The simple act of diving generates biophysical changes related to the loss of gravity and progress in a fluid (resistance to propulsion) or to thermal stress, with a drop in body temperature due to the enormous conductivity of water (3500 times higher than that of the air). The compensation manoeuvres used to balance the different sectors are also very important, with considerable differences in terms of pulmonary and circulatory pressure between the methods (classic manoeuvre of Valsalva or Valsalva "gentle", also known as Manovra di Marcante-Odaglia).

12.2.1.2 What Happens During Immersion?

Immersion is characterized by working in an extraordinary environment, which generates physical and physiological changes. Every 10 m of descent in seawater, there is an increase in environmental pressure of 100 kPa, equivalent to 1 atmosphere (1 bar). At a constant temperature (T), the volume (V) of a gas contained in a cavity (thorax, paranasal sinuses, middle ear and intestines) changes in an inversely proportional way with respect to pressure (P), with compression of the gas during the descent and expansion during the ascent (Boyle's law: $P \times V =$ constant). With the use of the self-breathing apparatus, the volumes remain constant if breathing is regular and without pauses, because the pressure balances inside and outside the rib cage, and it is guaranteed by the regulator. If the air inspired underwater is not expelled during the ascent, over-distension with pulmonary barotrauma (PB) and arterial gas embolism (AGE) (Boyle's law) can result. It is very important to keep in mind that the human body is an almost incompressible fluid.

Therefore, a barotrauma occurs when there is gas in inner rigid spaces and ambient pressure changes. During breathing, the gas level changes according to other physical laws. The partial pressure of gas increases proportionally to the increase in the ambient pressure (Dalton's law), and the amount of gas that is soluble in a liquid, at a given temperature, depends on the partial pressure of gas, as well as on its coefficient of solubility (Henry's law). Gases dissolve in fluids proportionately to their partial pressures in the gas phase, which means that the volume of dissolved gases in blood and tissues increases until equilibrium with the gas phase is achieved. The fluid is then saturated with the gas. During the compression and decompression phases of a dive, this process follows first-order exponential functions, since each tissue has its characteristic time constant. A greater amount of inert gas, especially nitrogen, solubilizes in the deep tissues and is removed during the ascent with the formation of bubbles, even in correct diving. The lung normally captures and removes from circulation the complex gas bubbles/proteins/platelets formed in the tissues ("filter function"), which is carried by the blood to the right atrium and to the pulmonary circulation. Only if an overproduction of bubbles cannot be eliminated does decompression sickness (DCS) occur. The presence of intrapulmonary shunts (congenital heart diseases with a right-left shunt, such as patent foramen ovale, previous pulmonary embolisms and arteriovenous malformations) increases the risk of DCS. The nitrogen distribution among the various compartments of the organism is characterized by a latency that, at the end of the submersion, leads to a nitrogen excess in the body with respect to pre-diving levels. Tissue supersaturation with inert gas (such as nitrogen or helium) during decompression is the mechanism of DCS. If the ascent is too fast in relation to the quantities of nitrogen absorbed, the solubilized nitrogen will be released during the gaseous phase through bubble formation in blood and tissues. Breathing during diving is also more difficult because the depth increases the density of inhaled gas (due to the increase of the number of molecules in a volume proportionally to the absolute pressure). Kinematic viscosity decreases, and there is therefore the possibility of greater turbulence. Laboured breathing during diving is due to a higher gas density and augmented resistances inside the pipes, regulator stages and the mouthpiece, both leading to an increase of the dead space. Respiratory work increases because of the combination of a greater density of gas and of hydrostatic pressure that modify the respiratory mechanics. Therefore, negative pressure is needed while inhaling, whereas a greater positive pressure is necessary while exhaling. The adult's maximum voluntary ventilation (MVV) in water is greatly reduced, from 200 L/min at the surface to 65 L/min at 50 m. This fact is due to the increases in gas density and airway resistance, as well as the redistribution of the blood volume from the periphery towards the centre (thorax), equal to approximately 700 mL, with a parallel decrease in pulmonary volume (blood shift). Finally, the relationship between ventilation and perfusion is altered with an inhomogeneous gaseous pulmonary exchange and a reduction of the maximum respiratory capacity. A reduction of the MVV, already present on the surface, will become even more marked during immersion, creating serious problems to the diver, who may need to perform a medium-high level of activity in emergency situations.

While exercising, a diver using negative-pressure breathing can seriously risk interstitial pulmonary oedema. The development of immersion pulmonary oedema is closely related to haemodynamic changes. The exercise-induced increase in tidal volume during immersion elevates the right heart preload, triggering a right-to-left ventricular imbalance and lung congestion. Exercising with negative-pressure breathing further increases the inspiratory work of breathing, right ventricle loading, right-to-left heart imbalance and rate of interstitial lung water accumulation. Positive-pressure breathing decreases cardiovascular changes and pulmonary oedema during immersion with exercise. Plasma atrial natriuretic peptide increases with inspiratory work and correlates with lung comet score (see Sec. 12.3). An altered right-to-left heart imbalance provokes the development of immersion pulmonary oedema when inspiratory work is high, e.g. during swimming at a high intensity or SCUBA diving with negative-pressure breathing.

When breathing air, the practical limit of pressure corresponds to a depth of 50–60 m of seawater (msw). The narcotic effects of nitrogen, the high gas density and the decompression stress contribute to adverse effects on the body beyond that depth range. Some of these effects can be reduced by changing the breathing gas mixture. Mixtures enriched with oxygen reduce the uptake of inert gas and thereby reduce the risks of decompression, whereas mixtures using added helium reduce the gas density. Increasing the oxygen fraction in the gas mixture amplifies the risk of toxic oxygen effects. Most dangerous are the risks of cerebral oxygen toxicity with sudden unconsciousness and seizures that can occur with exposure to partial pressures of oxygen higher than 150 kPa. Because of the risk of hyperoxic cerebral toxicity, military combat divers, who dive using pure oxygen in a closed-circuit breathing apparatus, usually restrict their diving depth range to 0–10 msw [47–53].

12.2.2 Fitness to Dive

Recreational free/breath-hold and SCUBA diving is a growing sport across the globe. It is evident that to prevent and minimize underwater accidents, the screening of the SCUBA divers is fundamental, with assessment of the physical and clinical conditions, especially respiratory, which can be conditioned and/or interfere with the immersion activity. The underwater candidate must not present:

- Anatomical or functional abnormalities that increase the risk of PB
- Impairment of the pulmonary function resulting in reduced operating capacity (underwater or on the surface)
- Limitation of lung function as a filter of the bubbles normally produced during the dive

In the past, asthma permanently and absolutely precluded underwater activity. More recently, a reversal trend has occurred, based on epidemiological studies that have correlated the theoretical risk to the actual risk, as well as to an evidence-based evaluation method [54–56].

In 2003, the British Thoracic Society (BTS) drew up guidelines on the respiratory aspects of fitness to dive, based on a reasoned revision of the literature and on the consensus of a group of experts. The consensus among diving experts worldwide is that subjects with a wheeze triggered by exercise, cold air or emotion should be advised not to dive [57–59]. In Australia, all divers with asthma must pass spirometry before certification, but in the United Kingdom, well-controlled asthmatics (excluding cold-, exercise- or emotion-induced asthmatics) may dive as long as they do not require a bronchodilator within 48 h. Among experts and other major diving organizations, the consensus is that lung function must be normal before an asthmatic can dive. Carefully selected mild to moderate, well-controlled asthmatics with normal screening spirometry can be considered candidates for diving per recommendations by the Recreational Scuba Training Council and the Undersea and Hyperbaric Medical Society (UHMS). Spirometry should be normal before and after exercise testing. Medication used to maintain normal spirometry is not a contraindication to diving. Inhalation challenge tests, including methacholine or hypertonic saline, are not recommended [60–64].

The presence of chronic obstructive pulmonary disease (COPD) contraindicates diving or requires diagnostic tests, due to a theoretical increase in the risk of barotrauma, in particular due to the possible presence of emphysema areas, and due to the reduction of exercise tolerance.

If the pulmonary function is reduced ($FEV_1/VC < 88\%$ of the theoretical with $FEV_1 < 80\%$ of the theoretical), the subject will be advised not to dive (Fig.12.2).

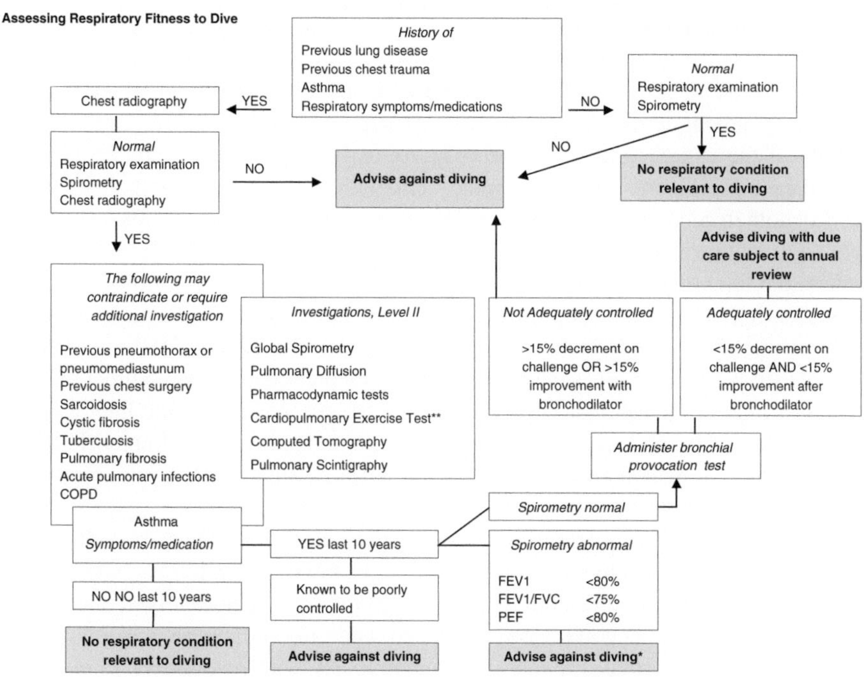

Fig. 12.2 Flow chart for assessing respiratory fitness to dive. (* Consider optimizing treatment and retesting; ** a peak capacity of 11 to 12 METS could be an appropriate goal)

The diver must have a sufficient aerobic capacity and be able to cope with increased respiratory work at depth, which can be exacerbated by exercise. In fact, a normal recreational dive does not require strenuous exertion, and a standard dive only needs a modest physical performance (3 METs, metabolic equivalents). However, in the event of an emergency (caused by equipment failures, currents, medical problems or other elements), intense physical exertion (7–10 METs) is required, and the diver must be able to sustain it. Others have suggested that a peak capacity of 11 to 12 METS could be an appropriate goal for recreational divers. This energy cost is estimated and does not consider variables such as body mass, adiposity, age, sex, movement efficiency and geographical and environmental conditions [65, 66].

In any case, in light of current knowledge, the final judgement can only be individualized. Diving is very safe if done properly, as there is minimal risk. However, if performed improperly, there is a risk of decompression illness and AGE. A preventive and periodic medical check, which includes spirometry, is the best guarantee to be able to carry out any underwater activity in safety, allowing for knowledge of the personal situation of the candidate and suggesting techniques and methods of immersion that are adapted to the subject.

12.3 Interstitial Lung Oedema

Interstitial oedema is the first appearance of water accumulation in the lung and affects a large majority of otherwise healthy climbers during acute exposure to high altitude. A rise in extravascular lung water may occur because of an increase in the pressure gradient across the microvascular barrier and/or by an increase in permeability of the endothelial barrier [67, 68]. Actually, hypoxia exposure, in addition to increasing pressures in the pulmonary circulation, increases endothelial permeability both in cell culture and in animal models. Interstitial lung oedema has a small but significant impact on the mechanical properties of lungs. The development of interstitial lung oedema (or subclinical lung oedema) during high-altitude exposure has been hypothesized a long time ago and was first inferred by some changes in spirometry, mainly the reduction of vital capacity (see Sec 12.1.4). We have to stress that the study of the changes in some lung function parameters is an indirect method to assess interstitial oedema. The gold standard to assess extravascular lung water is the imaging technique, especially high-resolution computed tomography (HRCT), which is more sensitive and specific than chest X-ray, but no HRCT studies at high altitude are available.

In the past, the assessment of the lung by echography has always been the main thing considered for evaluation of pleural effusion, but recently, ultrasound has been used to assess extravascular lung water [69]. In the case of normally aerated lung, the ultrasound beam is completely dissipated, and only the pleura is visualized as a hyperechoic horizontal line moving synchronously with respiration. In all cases where air content decreases and lung density increases (e.g. water, fibrosis), the acoustic mismatch between the lung and the surrounding tissues is lowered, and the ultrasound beam can be partly reflected at deeper zones and repeatedly. In this case, using the ultrasound probe, it is possible to visualize some vertical reverberation artefacts known as B lines. The ultrasound chest examinations can be performed using different probes (cardiac, linear or convex probe). Generally, a total of 28 echographic windows are scanned on each anterior and lateral hemithorax, and the presence of water or fibrosis in the interstitial space is identified by ultrasound B lines defined as echogenic, coherent, wedge-shaped signals with a narrow origin from the hyperechogenic pleural line. When a "white lung" pattern is observed in an intercostal space, an arbitrary value of 10 B lines is assigned for that space. The sum of the number of B lines measured at each of the 28 sites is calculated, and a score is obtained: ≤5 = normal; 5–15 = mild; 15–30 = moderate; >30 = severe [70]. Lung echo is used extensively in-hospital patients, in particular in intensive care units or in emergency departments, for the diagnosis of cardiogenic pulmonary oedema. Thanks to the diffusion of the portable echo machine, this method can be used easily in outpatients or in a wild environment.

Some years ago, a group of recreational trekkers performing a high-altitude ascent in Nepal were studied with lung and cardiac echography. They showed a high prevalence of clinically silent ultrasound B lines [71]. As matter of fact, lung ultrasound B lines were present in 83% of subjects starting from 3440 m and in 100% of subjects at 5130 m in the presence of normal left and right ventricular function. Furthermore, lung ultrasound B lines were mirrored by decreased oxygen saturation, whereas no statistically significant correlation with systolic pulmonary arterial pressure rise during ascent was observed. This study showed that the method was feasible in extreme environments and confirmed the hypothesis of the development of subclinical lung oedema during high-altitude exposure. This evidence was confirmed more recently by other groups [72, 73]. In particular, Pagé confirmed the development of extravascular lung water during high-altitude exposure independently of such confounding factors as previous acclimatization and physical activity. Subjects enrolled ascent rapidly to 4350 m by helicopter and were studied for 4 consecutive days at the same altitude. The B lines occurred early after the first morning, and a higher B-line number collated with maximum intensity of acute mountain sickness symptoms as assessed by Lake Louise score.

Like healthy subjects exposed acutely to high altitude, athletes performing heavy exercise at sea level can develop subclinical pulmonary oedema. There are numerous reports of pulmonary oedema using different techniques after prolonged exercise [74, 75]. During heavy exercise, there is an increase in cardiac output and a consequent increase in venous return associated with pulmonary blood flow heterogeneity, which causes regional overperfusion in the pulmonary vasculature, resulting in a marked increase in pulmonary capillary pressure. Most likely, this increase in pulmonary capillary pressure promotes fluid movement from the pulmonary vasculature to the interstitium in association with alveolar-capillary membrane leakage. Moreover, in elite athletes, when engaged in strenuous exercise, an arterial desaturation can be present that can be another possible cause of interstitial pulmonary oedema development during exercise [14, 76]. Lung echo can be used easily to investigate the development of extravascular lung water in athletes either in the laboratory or outdoors, before and after the performance. Few studies have used lung ultrasound to assess the presence of lung oedema after strenuous exercise. Pingitore et al. studied athletes performing an Ironman® triathlon distance (3.8 km swim, followed by a 180 km bike ride and a 42 km run) and found B lines in 74% of cases at the end of the race [77]. Multiple factor analysis showed a significant correlation between B lines and cardiac-related variables, and N-terminal pro-BNP B lines were associated with haemodynamic changes but not with the inflammatory response occurring during heavy and prolonged exercise. This study confirmed the "postcapillary" hypothesis regarding the development of extravascular lung water after intense and prolonged exercise. Certainly, more studies enrolling higher numbers of subject are needed to investigate better the pathophysiological roots of the development of extravascular lung water in healthy athletes.

Key Points

Altitude

- Ascent to high altitude leads to a progressive decrease in the PO_2 along the oxygen transport cascade.
- Individuals need to adapt to the progressive hypoxia (acclimatization process).
- The lung plays a pivotal role in the acclimatization process but may also be involved in maladaptive responses. The physiologic responses are the hypoxic ventilatory response and hypoxic pulmonary vasoconstriction. The maladaptive response is the development of high-altitude pulmonary oedema (HAPE).

Diving

- Immersion is characterized by working in an extraordinary environment: every 10 m of descent in seawater, there is an increase in environmental pressure of 100 kPa.
- At a constant temperature, the volume of a gas contained in a cavity (thorax, paranasal sinuses, middle ear and intestines) changes in an inversely proportional way with respect to pressure with compression of the gas during the descent and expansion during the ascent.
- Respiratory work increases because of the combination of a greater density of gas and of hydrostatic pressure modify the respiratory mechanics. The adult's maximum voluntary ventilation in water is greatly reduced.

Interstitial oedema

- Interstitial oedema is the first appearance of water accumulation in the lung and affects a large majority of otherwise healthy climbers during acute exposure to high altitude.

- The extravascular lung fluid accumulation at altitude can probably be considered a "paraphysiological" condition not necessarily proceeding to severe oedema with alveolar flooding.
- Like healthy subjects exposed acutely to high altitude, athletes performing heavy exercise at sea level can develop subclinical pulmonary oedema.
- Lung echo can be used easily to investigate the development of extravascular lung water.

References

1. Bärtsch P, Saltin B. General introduction to altitude adaptation and mountain sickness. Scand J Med Sci Sports. 2008;18(Suppl 1):1–10. https://doi.org/10.1111/j.1600-0838.2008.00827.
2. Pollard A, Murdoch D. The high altitude medicine handbook. Abingdon: Radcliff Medical Press; 1997. p. 1–2.
3. West JB. Highest permanent human habitation. High Alt Med Biol. 2002;3(4):401–7.
4. Pomidori L, Bonardi D, Campigotto F, Fasano V, Gennari A, Valli G, Palange P, Cogo A. The hypoxic profile during trekking to the pyramid laboratory. High Alt Med Biol. 2009;10(3):233–7. https://doi.org/10.1089/ham.2008.1107.
5. Burtscher M, Flatz M, Faulhaber M. Prediction of susceptibility to acute mountain sickness by SaO2 values during short-term exposure to hypoxia. High Alt Med Biol. 2004;5(3):335–40.
6. Lorente-Aznar T, Perez-Aguilar G, García-Espot A, Benabarre-Ciria S, Mendia-Gorostidi JL, Dols Alonso D, Blasco-Romero J. Estimation of arterial oxygen saturation in relation to altitude. Med Clin (Barc). 2016;147(10):435–40. https://doi.org/10.1016/j.medcli.2016.07.025.
7. West JB, Schoene RB, Luks AM, James S. Milledge high altitude medicine and physiology 5E. London: CRC Press; 2012.
8. Norton EF. The fight for Everest. London: Edward Arnold; 1925.
9. Messner R. Everest: expedition to the ultimate, baton wicks. London: Baton Wicks; 1979.
10. Bernardi L, Schneider A, Pomidori L, Paolucci E, Cogo A. Hypoxic ventilatory response in successful extreme altitude climbers. Eur Respir J. 2006;27(1):165–71.
11. Puthon L, Bouzat P, Rupp T, Robach P, Favre-Juvin A, Verges S. Physiological characteristics of elite high-altitude climbers. Scand J Med Sci Sports. 2016;26(9):1052–9. https://doi.org/10.1111/sms.12547.
12. Bernardi L, Passino C, Spadacini G, Bonfichi M, Arcaini L, Malcovati L, Bandinelli G, Schneider A, Keyl C, Feil P, Greene RE, Bernasconi C. Reduced hypoxic ventilatory response with preserved blood oxygenation in yoga trainees and Himalayan Buddhist monks at altitude: evidence of a different adaptive strategy? Eur J Appl Physiol. 2007;99(5):511–8.
13. Bilo G, Revera M, Bussotti M, Bonacina D, Styczkiewicz K, Caldara G, Giglio A, Faini A, Giuliano A, Lombardi C, Kawecka-Jaszcz K, Mancia G, Agostoni P, Parati G. Effects of slow deep breathing at high altitude on oxygen saturation, pulmonary and systemic hemodynamics. PLoS One. 2012;7(11):e49074. https://doi.org/10.1371/journal.pone.0049074.
14. Bernardi E, Pratali L, Mandolesi G, Spiridonova M, Roi GS, Cogo A. Thoraco-abdominal coordination and performance during uphill running at altitude. PLoS One. 2017;12(3):e0174927. https://doi.org/10.1371/journal.pone.0174927.
15. Cogo A, Legnani D, Allegra L. Respiratory function at different altitudes. Respiration. 1997;64(6):416–21.
16. Mason NP, Barry PW, Pollard AJ, Collier DJ, Taub NA, Miller MR, Milledge JS. Serial changes in spirometry during an ascent to 5,300 m in the Nepalese Himalayas. High Alt Med Biol. 2000;1(3):185–95.

17. Jaeger JJ, Sylvester JT, Cymerman A, Berberich JJ, Denniston JC, Maher JT. Evidence for increased intrathoracic fluid volume in man at high altitude. J Appl Physiol Respir Environ Exerc Physiol. 1979;47(4):670–6.
18. Mansell A, Powles A, Sutton J. Changes in pulmonary PV characteristics of human subjects at an altitude of 5,366 m. J Appl Physiol Respir Environ Exerc Physiol. 1980;49(1):79–83.
19. Contoli M, Bousquet J, Fabbri LM, Magnussen H, Rabe KF, Siafakas NM, Hamid Q, Kraft M. The small airways and distal lung compartment. Allergy. 2010;65:141–51. https://doi.org/10.1111/j.1398-9995.2009.02242.x.
20. Cremona G, Asnaghi R, Baderna P, Brunetto A, Brutsaert T, Cavallaro C, Clark TM, Cogo A, Donis R, Lanfranchi P, Luks A, Novello N, Panzetta S, Perini L, Putnam M, Spagnolatti L, Wagner H, Wagner PD. Pulmonary extravascular fluid accumulation in recreational climbers: a prospective study. Lancet. 2002;359:303–9.
21. Senn O, Clarenbach CF, Fischler M, Thalmann R, Brunner-La Rocca H, Egger P, Maggiorini M, Bloch KE. Do changes in lung function predict high-altitude pulmonary edema at an early stage? Med Sci Sports Exerc. 2006;38(9):1565–70.
22. Pellegrino R, Pompilio P, Quaranta M, Aliverti A, Kayser B, Miserocchi G, Fasano V, Cogo A, Milanese M, Cornara G, Brusasco V, Dellacà R. Airway responses to methacholine and exercise at high altitude in healthy lowlanders. J Appl Physiol (1985). 2010;108(2):256–65. https://doi.org/10.1152/japplphysiol.00677.2009.
23. Dellacà RL, Zannin E, Sancini G, Rivolta I, Leone BE, Pedotti A, Miserocchi G. Changes in the mechanical properties of the respiratory system during the development of interstitial lung edema. Respir Res. 2008;9:51. https://doi.org/10.1186/1465-9921-9-51.
24. Johnson MK, Birch M, Carter R, Kinsella J, Stevenson RD. Use of reactance to estimate transpulmonary resistance. Eur Respir J. 2005;25:1061–9.
25. Cogo A, Miserocchi G. Pro: most climbers develop subclinical pulmonary interstitial edema. High Alt Med Biol. 2011;12(2):121–4. https://doi.org/10.1089/ham.2011.0004.
26. Maggiorini M. High altitude-induced pulmonary oedema. Cardiovasc Res. 2006;72(1):41–50.. Epub 2006 Jul 12
27. Swenson ER, Bärtsch P. High-altitude pulmonary edema. Compr Physiol. 2012;2(4):2753–73. https://doi.org/10.1002/cphy.c100029.
28. Luks AM. Physiology in medicine: a physiologic approach to prevention and treatment of acute high-altitude illnesses. J Appl Physiol (1985). 2015;118(5):509–19. https://doi.org/10.1152/japplphysiol.00955.2014.
29. Luks AM, McIntosh SE, Grissom CK, Auerbach PS, Rodway GW, Schoene RB, Zafren K, Hackett PH, Wilderness Medical Society. Wilderness medical society consensus guidelines for the prevention and treatment of acute altitude illness. Wilderness Environ Med. 2010;21(2):146–55. https://doi.org/10.1016/j.wem.2010.03.002.
30. Stream JO, Luks AM, Grissom CK. Lung disease at high altitude. Expert Rev Respir Med. 2009;3(6):635–50. https://doi.org/10.1586/ers.09.51.
31. Seys SF, Daenen M, Dilissen E, Van Thienen R, Bullens DM, Hespel P, Dupont LJ. Effects of high altitude and cold air exposure on airway inflammation in patients with asthma. Thorax. 2013;68(10):906–13. https://doi.org/10.1136/thoraxjnl-2013-203280.
32. Bersuch E, Gräf F, Renner ED, Jung A, Traidl-Hoffmann C, Lauener R, Roduit C. Lung function improvement and airways inflammation reduction in asthmatic children after a rehabilitation program at moderate altitude. Pediatr Allergy Immunol. 2017;28(8):768–75. https://doi.org/10.1111/pai.12808.
33. Cogo A, Basnyat B, Legnani D, Allegra L. Bronchial asthma and airway hyperresponsiveness at high altitude. Respiration. 1997;64(6):444–9.
34. Huismans HK, Douma WR, Kerstjens HA, Renkema TE. Asthma in patients climbing to high and extreme altitudes in the Tibetan Everest region. J Asthma. 2010;47(6):614–9. https://doi.org/10.3109/02770900903573277.
35. Stokes S, Kalson N, Earl M, Frost H, Whitehead AG, Tyrrell-Marsh I, Davies A. Bronchial asthma on Mount Kilimanjaro is not a disadvantage. Thorax. 2008;63(10):936–7. https://doi.org/10.1136/thx.2008.102327.

36. Wu TY, Ding SQ, Liu JL, Yu MT, Jia JH, Chai ZC, Dai RC, Zhang SL, Li BY, Pan L, Liang BZ, Zhao JZ, Qi de T, Sun YF, Kayser B. Who should not go high: chronic disease and work at altitude during construction of the Qinghai-Tibet railroad. High Alt Med Biol. 2007;8(2):88–107.
37. Golan Y, Onn A, Villa Y, Avidor Y, Kivity S, Berger SA, Shapira I, Levo Y, Giladi M. Asthma in adventure travelers: a prospective study evaluating the occurrence and risk factors for acute exacerbations. Arch Intern Med. 2002;162(21):2421–6.
38. Cogo A, Fiorenzano G. Bronchial asthma: advice for patients traveling to high altitude. High Alt Med Biol. 2009;10(2):117–21. https://doi.org/10.1089/ham.2009.1007.
39. Dillard TA, Moores LK, Bilello KL, Phillips YY. The preflight evaluation. A comparison of the hypoxia inhalation test with hypobaric exposure. Chest. 1995;107(2):352–7.
40. Edvardsen A, Akerø A, Christensen CC, Ryg M, Skjønsberg OH. Air travel and chronic obstructive pulmonary disease: a new algorithm for pre-flight evaluation. Thorax. 2012;67(11):964–9. https://doi.org/10.1136/thoraxjnl-2012-201855.
41. Kelly PT, Swanney MP, Stanton JD, Frampton C, Peters MJ, Beckert LE. Resting and exercise response to altitude in patients with chronic obstructive pulmonary disease. Aviat Space Environ Med. 2009;80(2):102–7.
42. Furian M, Hartmann SE, Latshang TD, Flueck D, Murer C, Scheiwiller PM, Osmonov B, Ulrich S, Kohler M, Poulin MJ, Bloch KE. Exercise performance of lowlanders with COPD at 2,590 m: data from a randomized trial. Respiration. 2018;95(6):422–32. https://doi.org/10.1159/000486450.
43. Rand S, Prasad SA. Exercise as part of a cystic fibrosis therapeutic routine. Expert Rev Respir Med. 2012;6(3):341–51.; quiz 352. https://doi.org/10.1586/ers.12.19.
44. Fischer R, Lang SM, Brückner K, Hoyer HX, Meyer S, Griese M, Huber RM. Lung function in adults with cystic fibrosis at altitude: impact on air travel. Eur Respir J. 2005;25(4): 718–24.
45. Blau H, Mussaffi-Georgy H, Fink G, Kaye C, Szeinberg A, Spitzer SA, Yahav J. Effects of an intensive 4-week summer camp on cystic fibrosis: pulmonary function, exercise tolerance, and nutrition. Chest. 2002;121(4):1117–22.
46. Ryujin DT, Mannebach SC, Samuelson WM, Marshall BC. Oxygen saturation in adult cystic fibrosis patients during exercise at high altitude. Pediatr Pulmonol. 2001;32(6):437–41.
47. Lynch JH, Bove AA. Diving medicine: a review of current evidence. J Am Board Fam Med. 2009;22:399–407. https://doi.org/10.3122/jabfm.2009.04.080099.
48. Pendergast DR, Moon RE, Krasney JJ, Held HE, Zamparo P. Human physiology in an aquatic environment. Compr Physiol. 2015;5(4):1705–50. https://doi.org/10.1002/cphy.c140018.
49. Tetzlaff K, Thorsen E. Breathing at depth: physiologic and clinical aspects of diving while breathing compressed gas. Clin Chest Med. 2005;26(3):355–80.
50. Pendergast DR, Lundgren CE. The underwater environment: cardiopulmonary, thermal, and energetic demands. J Appl Physiol. 2009;106(1):276–83. https://doi.org/10.1152/japplphysiol.90984.2008.
51. Bosco G, Rizzato A, Moon RE, Camporesi EM. Environmental physiology and diving medicine. Front Psychol. 2018;9:72. https://doi.org/10.3389/fpsyg.2018.00072.
52. Ricciardi L. Modificazioni dell'apparato respiratorio in immersione. In: Rossi A, Schiavon M, editors. Apparato respiratorio e Attivita' Subacquea. Editeam S.a.S.Gruppo Editoriale; 2000, p. 1–15.
53. Attivita' subacquea e apparato respiratorio in. Criteri di valutazione pneumologia per l'idoneita' all'attivita' subacquea a cura di Maurizio Casasco e Annalisa Cogo. Med Sport. 2010;63(Suppl. 1 al N. 2):83–91.
54. Mebane GY. The coincidence of asthma and morbility or mortality in recreational scuba divers involving United Status citizens and reported to DAN. In: Elliott DH, editor. Are asthmatics fit to dive? UHMS; 1996, p. 3–5.
55. Di Napoli PL, Rossi A. Controindicazioni pneumologiche all'attività sportiva subacquea. In: Rossi A. e Schiavon M., editors. Apparato respiratorio e attività subacquea. Editeam sas Gruppo Editoriale; 2000, p. 25–41.

56. Schiavon M, Rossi A. Terapia farmacologia pneumologica: indicazioni e controindicazioni. In: Rossi A. e Schiavon M., editors. Apparato Respiratorio e attività subacquea. Editeam sas Gruppo Editoriale; 2000, p. 43–52.
57. British Thoracic Society Fitness to Dive Group. British Thoracic Society guidelines on respiratory aspects of fitness for diving. Thorax. 2003;58(1):3–13.
58. Adir Y, Bove AA. Can asthmatic subjects dive? Eur Respir Rev. 2016;25(140):214–20. https://doi.org/10.1183/16000617.0006-2016.
59. Coop CA, Adams KE, Webb CN. SCUBA diving and asthma: clinical recommendations and safety. Clin Rev Allergy Immunol. 2016;50(1):18–22. https://doi.org/10.1007/s12016-015-8474-y.
60. South Pacific Underwater Medicine Society. Suggested assessment for the diver with asthma. SPUMS [Online]. 2014. www.spums.org.au/public-file-download/full-spums-medical
61. Committee, UK Sport Diving Medical. Asthma and diving. British Sub-Aqua Club. [Online] September 16, 2014. www.bsac.com/page.asp?section=1533&sectionTitle=Respiratory+condition
62. Elliott DH. Are asthmatics fit to dive? Kensington, MD: Undersea and Hyperbaric Medical Society; 1996.
63. Recreational Scuba Training Council (RSTC). Medical Statement: Guidelines for recreational scuba diver's physical examination. https://wrstc.com/standards-dowloads/10-MedicalGuidelines. Accessed 10 June 2018.
64. Ustrup AS, Ulrik CS. Are recreational SCUBA divers with asthma at increased risk? J Asthma. 2017;54(8):784–91. https://doi.org/10.1080/02770903.2016.1263861.
65. Bosco G, Paoli A, Camporesi EM. Aerobic demand and scuba diving: concerns about medical evaluation. Diving Hyperb Med. 2014;44(2):61–3.
66. Mitchell SJ, Bove A. Medical screening of recreational divers for cardiovascular disease: consensus discussion at the Divers Alert Network Fatality Workshop. Undersea Hyperb Med. 2011;38:289–96.
67. Ogawa S, Shreeniwas R, Brett J, Clauss M, Furie M, Stern DM. The effect of hypoxia on capillary endothelial cell function: modulation of barrier and coagulant function. Br J Haematol. 1990;75(4):517–24.
68. Parker JCX, Parker RE, Granger DN, Taylor AE. Vascular permeability and trans- vascular fluid and protein transport in the dog lung. Circ Res. 1981;48(4):549–61.
69. Picano E, Pellikka PA. Ultrasound of extravascular lung water: a new standard for pulmonary congestion. Eur Heart J. 2016;37(27):2097–104. https://doi.org/10.1093/eurheartj/ehw164.. Epub 2016 May 12
70. Gargani L, Volpicelli G. How I do it: lung ultrasound. Cardiovasc Ultrasound. 2014;12:25. https://doi.org/10.1186/1476-7120-12-25.
71. Pratali L, Cavana M, Sicari R, Picano E. Frequent subclinical high-altitude pulmonary edema detected by chest sonography as ultrasound lung comets in recreational climbers. Crit Care Med. 2010;38(9):1818–23. https://doi.org/10.1097/CCM.0b013e3181e8ae0e.
72. Pagé M, Sauvé C, Serri K, Pagé P, Yin Y, Schampaert E. Echocardiographic assessment of cardiac performance in response to high altitude and development of subclinical pulmonary edema in healthy climbers. Can J Cardiol. 2013;29(10):1277–84. https://doi.org/10.1016/j.cjca.2013.04.025.. Epub 2013 Jul 17
73. Bouzat P, Walther G, Rupp T, Doucende G, Payen JF, Levy P, Verges S. Time course of asymptomatic interstitial pulmonary oedema at high altitude. Respir Physiol Neurobiol. 2013;186(1):16–21. https://doi.org/10.1016/j.resp.2012.12.009.. Epub 2012 Dec 29
74. Zavorsky GS, Milne EN, Lavorini F, Rienzi JP, Cutrufello PT, Kumar SS, Pistolesi M. Small changes in lung function in runners with marathon-induced interstitial lung edema. Physiol Rep. 2014;2(6):e12056. https://doi.org/10.14814/phy2.12056.
75. Taylor BJ, Carlson AR, Miller AD, Johnson BD. Exercise-induced interstitial pulmonary edema at sea-level in young and old healthy humans. Respir Physiol Neurobiol. 2014;191:17–25. https://doi.org/10.1016/j.resp.2013.10.012.. Epub 2013 Nov 4

76. Nielsen HB. Arterial desaturation during exercise in man: implication for O_2 uptake and work capacity. Scand J Med Sci Sports. 2003;13(6):339–58.
77. Pingitore A, Garbella E, Piaggi P, Menicucci D, Frassi F, Lionetti V, Piarulli A, Catapano G, Lubrano V, Passera M, Di Bella G, Castagnini C, Pellegrini S, Metelli MR, Bedini R, Gemignani A, L'Abbate A. Early subclinical increase in pulmonary water content in athletes performing sustained heavy exercise at sea level: ultrasound lung comet-tail evidence. Am J Physiol Heart Circ Physiol. 2011;301(5):H2161–7. https://doi.org/10.1152/ajpheart.00388.2011.

Pharmacological Management in Elite Athletes

13

Ken Fitch

Abstract

Elite athletes, who participate in sports that involve endurance training, have a high prevalence of asthma, exercise-induced bronchoconstriction (EIB) and airway hyperresponsiveness (AHR). Such prevalence is even higher when their sport necessitates inhaling large minute volumes of polluted or cold air. Elite athletes should be managed according to international guideline recommendations. Asthma guidelines consider inhaled glucocorticoids (GC) to be the gold standard preventive medication commonly supplemented with a long-acting β-2 agonist (LABA), while a short-acting β-2 agonist (SABA) should be inhaled as rescue therapy and pre-exercise. But elite athletes must heed the World Anti-Doping Agency's (WADA's) Prohibited List, and failing to do so could result in significant sanctions. Although many drugs prescribed to manage asthma and related conditions are permitted in sport, others are not or may be prescribed with restrictions. The latter includes salbutamol, the only SABA currently permitted. Systemic GC and adrenaline are prohibited in sport but only in-competition. However, a procedure, termed therapeutic use exemptions (TUEs), allows prohibited drugs to be prescribed to elite athletes when there is genuine medical need. The unpredictable pharmacokinetics of salbutamol and GC has and will continue to cause problems for elite athletes and their medical advisors. These topics are discussed in detail with mention of pitfalls for athletes and some of the potential issues that may arise for those who manage elite asthmatic athletes.

K. Fitch (✉)
School of Human Sciences—Sport, Exercise and Health, University of Western Australia, Crawley, WA, Australia
e-mail: ken.fitch@uwa.edu.au

A. Cogo et al. (eds.), *Exercise and Sports Pulmonology*,
https://doi.org/10.1007/978-3-030-05258-4_13

13.1 General Overview of the Topic

Elite athletes with asthma, EIB and AHR are managed similarly to other asthmatics who exercise, but there is an important difference because they and their medical advisors must be fully aware of and heed WADA's Prohibited List ('the List'). This chapter focuses on how 'the List' impacts on some medications used by elite athletes with asthma and what restrictions it places on them, how drugs that are prohibited or restricted in sport may be prescribed for and administered to elite athletes and some of the pitfalls that can cause problems for them and their doctors.

13.2 Introduction

Some elite athletes were encouraged to participate in sport and exercise when young to benefit their asthma and in time achieved excellence in their sport, but a greater percentage develop asthma, EIB and AHR after some years of intense endurance training [1]. Hence for many elite endurance athletes, it could be termed an occupational condition [2]. Inspired air needs to be humidified and warmed. When athletes breathe large or very large minute volumes of the ambient air for extended periods while participating in endurance training, the body may be unable to adequately warm and humidify the inspired air. Drying and cooling of the airways can cause airway injury and result in EIB, AHR and asthma [3]. Another important factor is the quality of the inspired air. Cold and very cold air and airborne pollutants including particulate matter (PM) and ozone (O_3) that commonly have their origins from combustion engines as well as chloramines and other products from purification of swimming pools can further damage the airways when large minute volumes are inspired regularly [4]. In these circumstances, prevention of airway injury is preferred to attempting to minimise it through pharmacological means [5].

Asthma, EIB and AHR constitute the most common chronic medical condition encountered in Olympic athletes, both summer and winter. Triathlon, cycling, swimming, and cross-country skiing, Nordic combined and speed skating are the sports with the highest prevalence of asthma, EIB and AHR which can range from 12 to 25% of athletes competing in those sports [1]. All these sports involve endurance training and the need to breathe large volumes of air which is frequently cold or polluted, and many consider this to be the reason for the high prevalence.

Other chronic medical conditions such as cystic fibrosis (CF), chronic obstructive pulmonary disease (COPD) and interstitial lung disease (ILD) cause such significant limitations of airway function as to prevent almost all with these conditions from achieving elite status in sport. Nevertheless, exercise has an important role in reducing the morbidity and enhancing the physical lifestyle of all with these three conditions, especially those with CF [6].

Doping, the use of prohibited drugs or methods to enhance training and sports performance, remains one of the greatest scourges of elite sport in the twenty-first century. The WADA's primary role is the global oversight of the WADA Code [7] and the Prohibited List [8] which is updated annually. Although the majority of medications to prevent and treat asthma are not performance enhancing, elite athletes with asthma and their medical advisors must be cognizant of and always heed the WADA Code and Prohibited List.

13.3 Pharmacological Management of Elite Athletes

The management of elite athletes should be consistent with the principles advised by GINA's asthma guidelines with inhaled GC representing the gold standard, commonly in combination with a LABA, while recommending SABA to be inhaled pre-exercise and when necessary to manage an acute exacerbation of asthma [9]. Other drugs such as cysteine-leukotriene receptor antagonists (LTRAs), mast cell stabilisers and muscarinic receptor antagonists (MRA) may be necessary as supplementary to the above, while currently, the prescription of xanthines is uncommon. IgE selective monoclonal antibody drugs are rarely necessary in allergic elite athletes, but H_1 receptor antagonists for associated atopic nasal or dermatological conditions are. The primary use of adrenaline in elite athletes is for acute anaphylaxis which appears to be relatively common although not every athlete with anaphylaxis has associated asthma and AHR. The status of each of these drug categories is classified according to the 2018 'List' in Table 13.1 [8].

No further comment is necessary concerning the permitted drug categories, but β_2-agonists, glucocorticoids and adrenaline demand detailed explanations. However, before enlarging, the concept of therapeutic use exemptions (TUEs) must be explained.

Table 13.1 Current status in sport of drugs to manage asthma and related conditions

Category of drug	Example	Status
Cysteine-leukotriene receptor antagonists (LTRAs)	Montelukast	Permitted
Muscarinic receptor antagonists (MRAs)	Ipratropium	Permitted
Mast cell stabilisers	Cromoglicate	Permitted
IgE selective monoclonal antibodies	Omalizumab	Permitted
H_1 receptor antagonists	Loratadine	Permitted
Xanthines	Theophylline	Permitted
β_2 adrenoceptor agonists: (1) short/rapid-acting (SABA) (2) long-acting (LABA)	Salbutamol Formoterol	'Permitted'/prohibited 'Permitted'/prohibited
Glucocorticoids: (1) inhaled (2) systemic	Budesonide Prednisolone	Permitted Prohibited
Adrenergic stimulant	Adrenaline	Prohibited

13.4 Therapeutic Use Exemptions (TUEs)

The concept of TUEs has been available for athletes for more than 25 years, and its origin was the International Olympic Committee's Medical Commission (IOC-MC) [10]. There are four criteria that must be met to allow a TUE Committee (TUEC) to approve an application for a TUE.

1. The drug is necessary to treat an acute or chronic medical condition, and the athlete would experience a significant impairment to health if it were to be withheld.
2. The therapeutic use of the prohibited drug is highly unlikely to produce any additional enhancement of performance beyond what might be anticipated by a return to the athlete's normal state of health following the treatment.
3. There is no reasonable permitted therapeutic alternative to the use of the prohibited drug.
4. The necessity to administer the prohibited drug is not a consequence of the prior misuse of a prohibited drug or method.

The above criteria and extensive detail as to the mechanism of TUEs including a template form to apply for a TUE are contained in the International Standard for Therapeutic Use Exemptions (ISTUE) [11]. Also able to be sourced from WADA's website is 'Medical Information to Support Decisions by TUECs'. This is divided into 19 separate documents, 2 of which are relevant to this chapter: asthma [12] and anaphylaxis [13]. Practical information as to how to apply for a TUE will be explained for β_2-agonists, glucocorticoids and adrenaline.

13.4.1 β_2-Agonists

Oral β_2-agonists have been prohibited in sport since the 1970s, soon after they were initially marketed. This occurred without evidence that they enhanced exercise performance. But in 1992, Martineau and colleagues demonstrated that 16 mg of slow-release salbutamol was anabolic [14], while the vast majority of studies have demonstrated that inhaled β_2-agonists do not improve sports performance [15, 16]. Two recent studies one from Canada when 1600 μg of salbutamol was inhaled [17] and one from England [18] when both 800 and 1600 μg of salbutamol were inhaled that demonstrated no enhancement of exercise performance are significant as these doses are the maximum permitted by WADA in 12 h (800 μg) and 24 h (1600 μg).

13.4.1.1 'The List' States

All oral and injected preparations of β_2-agonists are prohibited in sport [8].

13.4.2 Short-Acting Beta-2 Agonists (SABA)

(a) *Salbutamol.* Only one SABA is permitted—salbutamol—but with qualifications and because it has posed many problems for elite athletes, it necessitates a detailed explanation. Currently, 'the List' states:

- Inhaled salbutamol is permitted to a maximum 1600 μg over 24 h in divided doses not to exceed 800 μg over 12 h starting from any dose. The presence in urine of salbutamol in excess of 1000 ng/mL is not consistent with therapeutic use of the substance and will be considered as an adverse analytical finding (AAF) unless the athlete proves, through a controlled pharmacokinetic study, that the abnormal result was the consequence of a therapeutic dose (by inhalation) up to the maximum dose indicated above [8].
- The dosing limit, of 1600 μg in 24 h introduced by the IOC in 1997, was used in studies to distinguish prohibited oral from permitted inhaled salbutamol. From those studies, the combination of a urinary concentration of 1000 ng/mL of salbutamol plus a complicated statistical equation derived from the ratio of the active (R) enantiomer to the inactive (S) enantiomer of salbutamol was established to distinguish oral from inhaled salbutamol [19]. However, WADA has never embraced the statistical equation but has relied solely on the urinary concentration to consider that an athlete has misused salbutamol. Since 2002, the author is aware of the details of more than 20 athletes who have exceeded the urinary threshold of 1000 ng/mL, and all were in-competition. Other athletes have also exceeded the threshold, but in not one instance, there has been any evidence that an athlete has administered oral salbutamol. In the majority of cases, an increased quantity of salbutamol was inhaled because the athlete was managing troublesome bronchoconstriction. A common denominator has been a large proportion of salbutamol was inhaled in a short period which tends to significantly increase the urinary concentration [20]. Another factor is the variable pharmacokinetics of salbutamol which was first noted in 1997 when hospital patients were reported to demonstrate widely variable urinary salbutamol concentrations that did not correlate with the administered dosage [20]. Soon after, GlaxoSmithKline administered 1200 μg of inhaled salbutamol in 1 dose to 15 subjects, and the resultant urinary salbutamol concentrations were markedly different. One subject's urine had a concentration of 3000 ng/mL, another was 1500 ng/mL, 3 others were greater than 1000 ng/mL and the other 10 were less than 1000 ng/mL [21]. Dehydrated athletes produce urine with a high specific gravity (SG), and this can increase the urinary salbutamol [22]. WADA has acknowledged this, and from March 2018, urinary SGs above 1.020 are to be converted down to that figure but will involve the decision limit (DL) for salbutamol and not the urinary concentration of the drug [23]. The current DL for salbutamol is 1200 ng/mL, i.e. there is a 'leeway' of 200 ng/mL before the athlete is considered to have committed an AAF. For example, from March 2018, in urine with a SG of 1.030, the DL for salbutamol will be 1800 ng/mL.

- The author's examination of the circumstances when athletes have exceeded the 1000 ng/mL threshold with many having been sanctioned allows him to make the following recommendations to prevent or minimise such problems:
  - Unfortunately, a number of athletes have exceeded the 1000 ng/mL threshold when compelled to use a large quantity of salbutamol because they were not inhaling a GC. Except for a few athletes who experience the very occasional episode of EIB only and can be managed solely with pre-exercise salbutamol or another SABA, virtually all other elite athletes should be maintained on an inhaled GC, and many benefit from an accompanying LABA [9].
  - Nebulised salbutamol is considered to be an acceptable method of inhaling the drug by WADA, but hopefully this will change soon. The message is that nebulizing the drug greatly increases the risk of exceeding the 1000 ng/mL threshold and does not provide any benefit over inhaling salbutamol via a metered dose aerosol (MDI) with a spacer [24].
  - Should an athlete find it necessary to inhale a quantity of salbutamol greater than WADA allows or inhales the permitted amount in a period much less than 12 h and he/she is subjected to a doping control test within 36 h, an application for a retroactive TUE should be submitted promptly and before the test result is known. Such applications should be approved by a TUEC [12].
  - WADA's recommendation to undertake a pharmacokinetic study if an athlete exceeds the threshold [8] has been demonstrated to be unhelpful in most instances, and there are several reasons. Firstly, the study must be undertaken by the athlete after a 'washout period', i.e. the athlete must be salbutamol-free at the commencement of the study which, in almost every instance, differs from when the athlete had exceeded WADA's urinary threshold. Secondly, the study must be conducted in a laboratory, i.e. the athlete is resting in air-conditioned environment. In contrast, in every instance to date, the athlete has exceeded the urinary threshold during exercise which is known to increase the rate of excretion of salbutamol, and often the athlete was competing in either hot and humid or very cold conditions. Finally the pharmacokinetics of salbutamol can be quite unpredictable, and athletes have exceeded the 1000 ng/mL threshold after inhaling less salbutamol than WADA allows. In such circumstances, a pharmacokinetic study will not assist the athlete and may actually weaken his/her defence [25].

(b) *Other SABA*. All other SABA are prohibited in sport and can only be inhaled if the athlete has a TUE for that SABA. Ninety percent of Olympic athletes who inhale a SABA administer salbutamol [24]. In elite athletes, by far the most commonly used prohibited SABA is terbutaline, and in 2016, 87 athletes were identified by WADA as having an AAF for this SABA which approximates the mean annual number for the last 5 years [26]. How many of these 87 athletes were sanctioned is unknown.

13.4.2.1 Criterion 3 of the ISTUE States

There is no reasonable permitted therapeutic alternative to the use of the prohibited drug [11].

This should prevent a TUEC from approving an application for a TUE for a SABA other than salbutamol. However, WADA has acknowledged that it should not dictate medical treatment and accepts that when an athlete has been inhaling a SABA other than salbutamol for some time, TUECs should approve such applications if they meet the stated requirements [12]. A successful TUE application to inhale a SABA other than salbutamol demands:

- A detailed medical history and examination which should include the athlete's previous and current management with the prohibited SABA.
- The results of recent investigations such as spirometry or peak flow readings and an asthma diary if available.
- Recent evidence of AHR with either a positive bronchodilator test or a positive bronchial provocation test which can be either a physiological challenge such as eucapnic voluntary hyperventilation (EVH) or exercise or a pharmacological challenge including mannitol, hypertonic saline and methacholine.

A TUE approved for a SABA is valid for 4 years after which updated evidence of AHR must be submitted [12].

13.4.3 Long-Acting β2 Agonists (LABA)

Only two LABA, formoterol and salmeterol, are permitted in sport with dosage restrictions.

(a) *Formoterol.* 'The List' states: 'Inhaled formoterol: maximum delivered dose of 54 μg over 24 h. The presence in urine of formoterol in excess of 40 ng/mL is not consistent with therapeutic use of the substance and will be considered as an AAF unless the athlete proves, through a controlled pharmacokinetic study, that the abnormal result was the consequence of a therapeutic dose (by inhalation) up to the maximum dose indicated above' [8].

- WADA accepts that around 75% of the administered dose of formoterol is delivered via either a turbuhaler or aeroliser, thus allowing a daily maximum dose of 72 μg which complies with the manufacturer's recommendation [13]. Currently, to the author's knowledge, no athlete has exceeded the urinary threshold of 40 ng/mL which is considered 'generous', and hence no athlete has needed to undergo a pharmacokinetic study inhaling formoterol.

(b) *Salmeterol.* 'The List' states: 'Inhaled salmeterol: maximum 200 μg over 24 h' [8], which is the manufacturer's recommended daily maximum dose.

13.4.4 Glucocorticoids (GC)

'The List' states: 'All glucocorticoids are prohibited when administered by oral, intravenous, intramuscular or rectal routes' in-competition only [8].

(a) *Inhaled.* All inhaled GC used to manage asthma are permitted in and out of competition. Only one aspect has caused problems—the detection of a metabolite of inhaled budesonide—16α-hydroxy-prednisolone—in urinary concentrations greater than the reported threshold for all synthetic GC of 30 ng/mL. Because oral budesonide which is prohibited in sport is used to treat inflammatory bowel disease, it is necessary to ensure that budesonide in an athlete's urine has resulted from inhaled and not oral administration. 16α-Hydroxy-prednisolone was frequently present in urine 7–10 times the concentration of the parent compound budesonide and well above 30 ng/mL threshold. That the concentration of budesonide was always below the threshold was able to negate any perceived AAFs from this metabolite. Recent research in Barcelona has identified another budesonide metabolite, 6β-hydroxy-budesonide, and that a urinary threshold of 20 ng/mL of 6β-hydroxy-budesonide would distinguish oral from inhaled budesonide [27]. Thus there is no longer any possible confusion about budesonide which is inhaled more often by Olympic athletes than any other GC.
(b) *Systemic.* Although intravenous GC is occasionally used for status asthmaticus in hospital settings, oral GC—mostly prednisolone—is the commonest prohibited substance for which Olympic athletes seek a TUE and acute exacerbations of asthma a frequent reason [10]. As oral GC is permitted out of competition, in many instances the drug can be prescribed for an elite athlete without a TUE or any concern for the prescribing doctor because either the athlete has no imminent competitions or is too ill with asthma to participate in his/her sport. However, experience has demonstrated that frequently, acute asthma occurs shortly prior to an event and this can pose significant difficulties. If the athlete's asthma responds well to an oral GC and it is necessary for that athlete to continue the drug on the competition day or later, a TUE must be obtained. Applying for a TUE necessitates submitting a detailed history and examination with spirometric evidence of increased airway obstruction that has not been controlled by full doses of an inhaled GC and the need for increased rescue therapy with a SABA.

A contentious problem can be if the oral GC is ceased prior to the competition, when will it no longer be detectable in the urine? This is a question that is difficult to answer with confidence in part because the pharmacokinetics of GC can be unpredictable. The best advice is that to ensure that GC is not detected in the urine and thus avoid an AAF, if an oral GC is ceased less than 4–5 days prior to an event, apply for a TUE for the oral GC.

13.4.5 Adrenaline (Epinephrine)

The use of adrenaline is primarily for the management of anaphylaxis, which is a serious, life-threatening, systemic hypersensitivity reaction that is rapid in onset and has a lifetime prevalence of 0.05–2%. The rate of occurrence appears to be increasing especially in young people. Common triggers include foods, insect venoms and drugs as well as exercise, cold and heat. That WADA has considered it necessary to establish a document to assist TUECs and discusses the requirements to grant a TUE for adrenaline for anaphylaxis [13] is evidence that elite athletes do experience this condition with some frequency. As the author chaired the IOC's and his nation's TUEC for more than 20 years, he can confirm this. Anaphylaxis is a clinical diagnosis not requiring laboratory confirmation. Acute anaphylaxis is a medical emergency that demands prompt treatment with an intramuscular injection of adrenaline in the mid-lateral thigh in a dose of 0.01 mg/kg of a 1:1000 (1 mg/1 mL) solution and a maximum of 0.5 mg for adolescents and adults. This may be repeated in 5–15 min if necessary. Commonly, 0.3 mg of adrenaline is packaged in an auto-injector for such use. As adrenaline is prohibited in-competition only, a retroactive TUE is required only if the injection is administered very close to an event. It is essential for all athletes who experience an episode of anaphylaxis to carry such a syringe of adrenaline at all times. As adrenaline is a prohibited substance, athletes clearing customs can encounter difficulties with overzealous officials, and it is prudent for these athletes to seek a prospective TUE which should be approved and valid for 5 years. The TUE approval certificate should be carried by the athlete and produced if necessary to avoid problems. Other treatments for anaphylaxis and associated secondary effects include inhaled salbutamol (see above), an oral non-sedating H_1-antihistamine (permitted) and systemic GC either orally or intravenously (prohibited). However, it is stressed that GC is only a second-line treatment for anaphylaxis and unlike adrenaline is not rapidly effective and should never be the sole treatment [13].

13.5 Conclusions

WADA Prohibited 'List' does pose significant problems for elite athletes with asthma. Although only one SABA, salbutamol, is permitted in sport, it has proved to be a major source of difficulty for both athletes and their medical advisors. These concern the urinary threshold of 1000 ng/mL for the drug and that this can be achieved because an athlete has an exacerbation of asthma requiring an increased amount of inhaled salbutamol as rescue therapy and because of the unpredictable pharmacokinetics of the drug. Other SABA such as terbutaline can only be prescribed for athletes with a TUE. Athletes and the doctors need to be constantly alert to the WADA List and any changes that occur annually and always heed the 'List'.

13.6 Tips and Pitfalls

- Ensure your elite athletes with asthma, AHR and EIB are appropriately medicated as guidelines advise and stress that they should endeavour not to overuse inhaled salbutamol.
- Regularly scrutinise your athlete-patients' inhalation technique, and remind them to clean their MDIs and spacers often.
- Counsel your elite athlete-patient about WADA restrictions on dosing with salbutamol and the possible consequences of inhaling larger than customary quantities in short periods of time.
- When young asthmatic athletes graduate to being classed as elite and thus are required to heed the WADA 'List', ensure that they are not administering any prohibited drugs and are aware of 'the List'. If they are inhaling, for example, terbutaline, determine whether he/she should switch to salbutamol or not and if not, immediately seek a TUE.
- Be aware that an elite athlete-patient who is not using an inhaled GC may be rejected for a TUE for oral prednisolone to manage an exacerbation of asthma because a permitted alternative—an inhaled GC—is available and should be tried first.
- For any athlete-patient with anaphylaxis, stress the necessity of carrying an auto-injector with them at all times and seek a prospective TUE to avoid issues with customs.
- Despite the above, remember the WADA advice 'an athlete's health should never be jeopardized by withholding medication in an emergency' [12].

Key Points

- The triad of asthma, exercise-induced bronchoconstriction (EIB) and airway hyperresponsiveness (AHR) constitutes the commonest medical condition encountered in Olympic athletes.
- Athletes with this triad should be managed in accordance with the Global Initiative for Asthma's (GINA's) recommendations with inhaled glucocorticoids as the preventive medication for almost all.
- But elite athletes and their medical advisors must be aware of and heed the World Anti-Doping Agency's (WADA's) prohibited list which is updated annually.
- Inhaled salbutamol, the short-acting β_2 agonist (SABA) used by the vast majority of athletes with the triad, has caused and can be expected to continue to cause more issues than all other asthma medications combined.
- Athletes with asthma and their medical advisors should be cognizant of the concept of therapeutic use exemptions (TUEs) which may need to be invoked notably when an oral glucocorticoid may be necessary to manage acute asthma.

Case Study

A 19-year-old female cross-country skier attends you for the first time about her troublesome asthma, concerned because she has been chosen to compete in her first Olympic Games, and her event, a 15 km race, is scheduled for the first competition day in 14 days. You learn that her asthma commenced when aged 17 soon after she had represented her country at the Winter Youth Olympic Games. She consulted her GP who prescribed salbutamol, 200 µg, to be inhaled via a MDI prior to exercise and whenever her asthma was troublesome. Since then, she has followed that advice but has not used more than 4 MDIs in either of the last 2 years. However, about a week ago, she developed the typical symptoms of an upper respiratory tract infection which soon involved her chest, and during the past 3 days, she has required as many as 12–14 (1200–1400 µg) puffs of salbutamol a day. You examine her and note that she is coughing and has a slight wheeze. Spirometry reveals her FEV_1 is 70% of predicted and her $FEV_{T\%}$ is 68%.

How would you manage her? (More than one answer may be correct)

A. Administer 200 µg of salbutamol and repeat her spirometry?
B. Prescribe 40 mg of oral prednisolone a day and ask her to return in 3 days?
C. Prescribe inhaled beclomethasone 200 µg twice a day and return in a week.
D. Prescribe 40 mg of oral prednisolone a day, complete an application form for a TUE and ask her to return in 5 days?
E. Prescribe a preparation of budesonide and formoterol 200/6–2 puffs to be inhaled twice a day and return in 3 days.
F. Counsel her about the dangers of inhaling large quantities of salbutamol

- Answer A correct. It would be advantageous to ascertain if she can provide evidence of a positive bronchodilator test and she does with her FEV_1 increasing 17% after inhaling 200 µg of salbutamol. A positive test is a 12% increase in FEV_1 or greater.
- Answer B is not the preferred treatment as she is GC naïve and should respond to the combined inhaled treatment.
- Answer C may be acceptable, but given she has a major event in only 14 days, the combination of a GC with a LABA is superior.
- Answer D is incorrect. Oral prednisolone is not the optimal first-line treatment, and no TUE is necessary 14 days out from competition. Furthermore, if an application was made, the TUEC should reject the application because a permitted alternative—inhaled GC—has not been tried first.
- Answer E correct. An inhaled glucocorticoid (GC) is overdue, and as she needs prompt improvement in her bronchoconstriction, a combination of a GC and a long-acting β-2 agonist (LABA) is the best treatment.
- Answer F correct. This is an excellent time to counsel her about salbutamol and how it can provoke an AAF.
- Answers A, E, and F are correct.

The skier reports back in 3 days and pleasingly is somewhat improved although not fully recovered, and the following day, the team flies out to the Olympic city. After settling into the Olympic Village, the skier returns to your medical centre very upset as she became breathless and wheezy during her first training session at the cross-country centre this morning. Her first race is now only a week away. You repeat her spirometry and note that her FEV_1 is 76% of predicted and $FEV_{T\%}$ is 73%. You check her inhalation technique and it appears to be satisfactory. The skier informs you that it was very cold when she was training today, and you speak with her coach and he advises that at the cross-country venue, this morning's temperature was −12 °C, much colder than at the national cross-country skiing training centre. You ask if the team has any heat and moisture exchange (HME) masks, and the coach informs you that they do. You request him to provide the skier with one for tomorrow's training session and advise her to continue using the combination inhaler twice daily, increase her inhaled salbutamol to 4 doses each of 200 μg spread over 14 h and report to you tomorrow immediately after training. She does and you learn that the mask did help but she still struggled for breath. Her spirometry is marginally improved from the previous day, and it is obvious that she cannot perform at the level necessary to participate in the games. The HME mask would not be suitable to be used in-competition but can reduce the potential airway injury that can occur when breathing large minute volumes of very cold, dry air while training.

How would you manage her?

G. As she has a valid bronchodilator test performed only 8 days earlier, commence treatment immediately with oral prednisolone 40 mg a day and apply for a therapeutic use exemption (TUE) from her International Federation's (FIS's) TUE Committee (TUEC) to administer it for 5 days.
H. As she is now 'in-competition', request an urgent TUE from the IOC's TUEC for oral prednisolone 40 mg a day with a reducing dose for 7 days and wait for it to be approved before commencing treatment.
I. Administer 250 mg of hydrocortisone sodium succinate intravenously followed by 40 mg prednisolone daily for 5 days and seek a TUE from the IOC's TUEC.
J. Commence 40 mg oral prednisolone immediately and seek a TUE from the IOC's TUEC for 40 mg oral prednisolone daily for 7 days.
  - G is incorrect. As the skier is now resident in the Olympic Village, she is an Olympic athlete and thus is bound by Olympic and not FIS rules. Hence, it is the IOC and not FIS who must receive the TUE application.
  - H is incorrect. Until the revised International Standard for TUEs (ISTUE) 2015 became valid, the IOC considered that 'in-competition' commenced when the Olympic Village opened. Hence, oral prednisolone was prohibited for the duration that the Village was open. But the

ISTUE 2015 defined 'in-competition' as being from 12 h prior to the commencement of an event so 6 days prior was out-of-competition, and as the IOC now heeds this rule, no TUE was necessary to commence oral prednisolone.

- I is incorrect. There is no justification for IV hydrocortisone, and if injected in more than 100 mL of fluid, it would necessitate an additional TUE for an IV infusion.
- J is the correct answer. As it is now 6 days prior to her competition, no TUE is necessary to commence the treatment. But as it is highly likely that oral prednisolone will be continued until the competition or if ceased pre-competition, evidence of its use will be detectable if tested post-event, a TUE should be sought from the IOC's TUEC. Her positive bronchodilation test performed 8 days earlier, inhaling a GC and a LABA and her spirometry data would suffice for the TUE to be granted. Many experts do not consider that 'taping off' a short course of oral GC is necessary, but if you do chose to do so, this could still be undertaken. In fact, a full 7-day course may be unnecessary, but when the TUE is approved, it remains available if required. The TUE will allow for prednisolone to be detected and reported in her urine if the skier is tested after her first event and would avoid any AAF.

13.7 Outcome

The skier responds dramatically to prednisolone, and she reports back to you after her final training session 2 days before her event and is feeling 'great'; her FEV_1 is 97% of predicted and her $FEV_{T\%}$ is a satisfactory 83%. You reduce her prednisolone to 20 mg for the following day, and as her post-training pulmonary function remains unchanged, prednisolone is ceased a day prior to her event. You reduce her salbutamol to 2 puffs (200 μg) pre-exercise. She competes in her 15 km race, finishing a credible 12th, and her time is only 2 s slower than her personal best. The skier is advised to continue using her budesonide/formoterol inhaler twice daily and to inhale 200 μg of salbutamol prior to exercise.

References

1. Fitch KD. An overview of asthma and airway hyper-responsiveness in Olympic athletes. Br J Sports Med. 2012;46:413–6.
2. Price OJ, Ansley L, Menzies-Gow A, Cullinan P, Hull JH. Airway dysfunction in elite athletes-an occupational lung disease? Allergy. 2013;68:1343–52.
3. Anderson SD, Kippelen P. Airway injury as a mechanism for exercise-induced bronchoconstriction in elite athletes. J Allergy Clin Immunol. 2008;122:225–35.

4. Fitch KD, Anderson SD. Intense exercise and airway hyper-responsiveness/asthma importance of environmental factors. Br J Sports Med. 2012;46:379–80.
5. Kippelen P, Fitch KD, Anderson SD, Bougault V, Boulet L-P, Rundell KW, et al. Respiratory health of elite athletes – preventing airway injury: a critical review. Br J Sports Med. 2012;46:471–6.
6. Radtke T, Nevitt SJ, Hebestreit H, Kriemler S. Physical exercise training for cystic fibrosis. Cochrane Database Syst Rev. 2017;11:CD002768.
7. WADA Code. https://www.wada-ama.org/en/what-we-do/the-code. Accessed 20 July 2018.
8. World Anti-Doping Agency (WADA) 2018 Prohibited List. https://www.wada-ama.org/sites/default/files/prohibited_list_2018_en.pdf. Accessed 11 Jan 2018.
9. The Global Asthma Report 2014. http://www.globalasthmareport.org/. Accessed 11 Jan 2018.
10. Fitch KD. Therapeutic use exemptions (TUEs) at the Olympic Games 1992–2012. Br J Sports Med. 2013;47:815–8.
11. World Anti-Doping Agency (WADA) 2015 ISTUE. https://www.wada-ama.org/sites/default/files/resources/files/WADA-2015-ISTUE-Final-EN.pdf. Accessed 12 Jan 2018.
12. World Anti-Doping Agency (WADA) Medical information to support decisions of TUECs – Asthma. https://www.wada-ama.org/sites/default/files/tpg_-_asthma_-_version_6.0_-_december_2017.pdf. Accessed 13 Jan 2018.
13. World Anti-Doping Agency (WADA) Medical information to support decisions of TUECs – Anaphylaxis. https://www.wada-ama.org/sites/default/files/resources/files/wada-tpg-anaphylaxis-2.1-en.pdf. Accessed 13 Jan 2018.
14. Martineau L, Horan MA, Rothwell NJ, Little RA. Salbutamol, a beta 2-adrenoceptor agonist, increases skeletal muscle strength in young men. Clin Sci (Lond). 1992;83:615–21.
15. Kindermann W. Do inhaled beta(2)-agonists have an ergogenic potential in non-asthmatic competitive athletes? Sports Med. 2007;37:95–102.
16. Pluim BM, de Hon O, Staal JB, Limpens J, Kuipers H, Overbeek SE, et al. β2-agonists and physical performance. Sports Med. 2011;41:39–57.
17. Koch S, Ahn JR, Koehle MS. High-dose inhaled salbutamol does not improve 10-km cycling time trial performance. Med Sci Sports Exerc. 2015;47:2373–9.
18. Dickinson J, Hu J, Chester N, Loosemore M, Whyte G. Acute impact of inhaled short acting b2-agonists on 5 km running performance. J Sports Sci Med. 2014;13:271–9.
19. Berges R, Segura J, Ventura R, Fitch KD, Morton AR, Farre M, et al. Discrimination of prohibited oral use of salbutamol from authorized inhaled asthma treatment. Clin Chem. 2000;46:1365–75.
20. Jacobson GA, Peterson GM, McLean S. Investigation of urinary levels of salbutamol in asthmatic patients receiving inhaled therapy. J Clin Pharm Ther. 1997;22:119–26.
21. GlaxoSmithKline. GSK Study: GM1998/00055/00. 1998.
22. Dickinson J, Hu J, Chester N, Loosemore M, Whyte G. Impact of ethnicity, gender, and dehydration on the urinary excretion of inhaled salbutamol with respect to doping control. Clin J Sport Med. 2014;24:482–9.
23. World Anti-Doping Agency – TD2018DL Decision Limits for the Confirmatory Quantification of Threshold Substances.https://www.wada-ama.org/en/resources/science-medicine/td2018dl. Accessed 13 Jan 2018.
24. Fitch K. The World Anti-Doping Code: can you have asthma and still be an elite athlete? Breathe. 2016;12:148–58.
25. Fitch KD. The enigma of inhaled salbutamol and sport: unresolved after 45 years. Drug Test Anal. 2017;9:977–82.
26. World Anti-Doping Agency – Anti-doping testing figures. https://www.wada-ama.org/sites/default/files/resources/files/2016_anti-doping_testing_figures.pdf. Accessed 13 Jan 2018.
27. Matabosch X, Pozo OJ, Pérez-Mañá C, Farré M, Marcos J, Segura J, et al. Discrimination of prohibited oral use from authorized inhaled treatment of budesonide in sports. Ther Drug Monit. 2013;35:118–28.

Environmental Conditions, Air Pollutants, and Airways

14

Giuseppe Morici, Fabio Cibella, Daniele Zangla, Pierpaolo Baiamonte, and Maria R. Bonsignore

Abstract

Air pollution is a major problem worldwide, which could be even more serious for athletes who train in urban environments. Exercise increases minute ventilation and exposure to pollutants, but the literature on the effects of air pollution in athletes is relatively scarce, with the exception of chlorine exposure in athletes of aquatic sports and air pollution secondary to ice resurfacing in athletes perform-

G. Morici
Dipartimento di Biomedicina Sperimentale e Neuroscienze Cliniche (BioNeC), Università di Palermo, Palermo, Italy

Istituto di Biomedicina e Immunologia Molecolare (IBIM), Consiglio Nazionale delle Ricerche (CNR), Palermo, Italy
e-mail: giuseppe.morici@unipa.it

F. Cibella
Istituto di Biomedicina e Immunologia Molecolare (IBIM), Consiglio Nazionale delle Ricerche (CNR), Palermo, Italy
e-mail: fabio.cibella@ibim.cnr.it

D. Zangla
Dipartimento di Scienze Psicologiche, Pedagogiche e della Formazione (SPPF), Università di Palermo, Palermo, Italy
e-mail: daniele.zangla@unipa.it

P. Baiamonte
Dipartimento di Biomedicina e Medicina Interna e Specialistica (diBiMIS), Università di Palermo, Palermo, Italy
e-mail: maiden59.mail@libero.it

M. R. Bonsignore (✉)
Istituto di Biomedicina e Immunologia Molecolare (IBIM), Consiglio Nazionale delle Ricerche (CNR), Palermo, Italy

Dipartimento di Biomedicina e Medicina Interna e Specialistica (diBiMIS), Università di Palermo, Palermo, Italy
e-mail: mariarosaria.bonsignore@unipa.it

A. Cogo et al. (eds.), *Exercise and Sports Pulmonology*,
https://doi.org/10.1007/978-3-030-05258-4_14

ing in ice arenas. Although air pollution may exert detrimental effects on athletic performance, little has been published on this topic. The largest body of information regards the impact of air pollution during urban active transport, i.e., walking and cycling in cities, due to the potential risk of air pollution in citizens and the need to rethink urban transportation strategies accordingly. In healthy subjects, the benefits of physical activity largely outweigh the disadvantages of exposure to air pollutants. In susceptible individuals, however, such as patients with cardiac or respiratory disease and children, detrimental effects have been demonstrated. Improvement in air quality, individual protective behaviors, and prompt communication to the population of dangerous air quality may help to limit the negative effects of air pollution on respiratory health.

14.1 Introduction

Air pollution is increasingly considered as a major public health issue, especially since current estimates foresee that about 70% of the world population will live in very extended urban environments by 2050 [1]. Such scenario means that urban design, air pollution control, as well as detailed plans for public and private transportation [2–4] will be necessary to preserve population health from increasing levels of air pollutants.

Environmental evaluation, however, is a very complicated issue. First, exposure to multiple pollutants can derive from different sources (e.g., vehicle traffic, industries, but also fireworks, forest fires, or volcanic eruptions), making data obtained during exposure to single pollutants not necessarily generalizable to real-world situations and requiring complex statistical modeling [5]. Second, the pattern of climate and winds differs among locations, thus modifying exposure settings. Third, monitoring of air quality is evolving, from fixed monitoring stations to wearable devices [6, 7] allowing more precise definition of individual exposures. Finally, susceptibility to the noxious effects of pollutants varies according to age and pre-existing cardiovascular and respiratory diseases. There is evidence that air pollution is associated with cardiorespiratory diseases [8], and COPD patients are particularly vulnerable to air pollutants, both in terms of mortality or hospital admission risk and pollution-associated decline in respiratory function [9].

At the national level, the Environmental Protection Agency (EPA) in the United States focused on reducing air pollution and produced the National Ambient Air Quality Standards for pollutants in order to protect citizens' health. EPA created the air quality index (AQI), an index summarizing air quality day by day. Thus, American citizens can easily access AQI on the web to understand the possible threats for their own health and to access local pollution data [10].

The 2015 WHO Consultation [11] stated that air pollution has to be recognized as a threat to human health, being associated with increase in both mortality and morbidity worldwide. Moreover, the effects of environmental pollution increased steadily in the last decades. Almost all the more common ambient pollutants produce

Table 14.1 Type, source, and respiratory effects of the most common air pollutants[a] [25]

Type	Source	Main respiratory effects
Ozone (O_3)	Photochemical reactions from vehicular traffic (secondary pollutant)	Reduction of lung function Increase of bronchial hyperresponsiveness Reduced exercise tolerance
Nitrogen dioxide (NO_2)	Each combustion process having atmospheric air as comburent	Increase of bronchial hyperresponsiveness Reduction of lung function Reduced exercise tolerance
Sulfur dioxide (SO_2)	Fuel combustion (from industry and vehicular traffic)	Reduction of lung function
Respirable fractions of particulate matter	Fuel combustion (from industry and vehicular traffic), nonindustrial combustion processes	Reduction of lung function
United Nations. 2014. *World Urbanization Prospects: The 2014 Revision, Highlights.* New York: United Nations, Department of Economic and Social Affairs, Population Division. http://esa.un.org/unpd/wup/Highlights/WUP2014-Highlights.pdf	United Nations. 2014. *World Urbanization Prospects: The 2014 Revision, Highlights.* New York: United Nations, Department of Economic and Social Affairs, Population Division. http://esa.un.org/unpd/wup/Highlights/WUP2014-Highlights.pdf	United Nations. 2014. *World Urbanization Prospects: The 2014 Revision, Highlights.* New York: United Nations, Department of Economic and Social Affairs, Population Division. http://esa.un.org/unpd/wup/Highlights/WUP2014-Highlights.pdf

[a]G. Viegi, S. Baldacci. Epidemiological studies of chronic respiratory conditions in relation to urban air pollution in adults. Eur Respir Mon 2002, 21, 1–16

respiratory effects (Table 14.1, Ref. [12]) in individuals, in particular in vulnerable categories (children, subjects affected by chronic respiratory disease such as asthma or chronic obstructive pulmonary disease, subjects affected by cardiovascular diseases, older adults). In its 2015 revision of air quality guideline [11], WHO experts identified 32 air pollutants divided in 4 groups. For most of them, recent evidence justifies re-evaluation: in particular, particulate matter (PM), ozone, nitric dioxide (NO_2), sulfur dioxide (SO_2), and carbon monoxide (CO) need systematic re-evaluation due to increased knowledge about their detrimental effects on health. Along the same line, the American Thoracic Society (ATS) has recently published an estimate of excess mortality and morbidity in the United States, including lung cancer, according to the ATS-recommended thresholds for $PM_{2.5}$ and ozone concentrations [13], which are lower than the currently recommended values (Table 14.2).

14.2 Air Pollution and Exercise

It is well established that physical exercise has important benefits on individual's health [14]. On the other hand, people living in urban or highly industrialized areas are exposed to high levels of environmental pollution. Recreational outdoor exercise

Table 14.2 Upper limit values for outdoor pollutants

Pollutant	Limit	Recent ATS recommendations [13]
$PM_{2.5}$[a]	10 μg/m³ annual mean 25 μg/m³ 24-h mean	≤11 μg/m³ annual mean ≤25 μg/m³ 24-h mean
PM_{10}[a]	20 μg/m³ annual mean 50 μg/m³ 24-h mean	
O_3[a]	100 μg/m³ 8-h mean	≤0.060 ppm 8-h mean
NO_2[a]	40 μg/m³ annual mean 200 μg/m³ 1-h mean	
SO_2[a]	20 μg/m³ 24-h mean 500 μg/m³ 10-min mean	
CO[b]	10 mg/m³ 8-h mean 40 mg/m³ 1-h mean	

[a]WHO Regional Office for Europe (2006). *Air quality guidelines global update 2005: particulate matter, ozone, nitrogen dioxide and sulfur dioxide*. Copenhagen, WHO Regional Office for Europe. Available at http://www.euro.who.int/__data/assets/pdf_file/0005/78638/E90038.pdf
[b]EPA—United States Environmental Protection Agency. National Ambient Air Quality Standards (NAAQS). Available at https://www.epa.gov/criteria-air-pollutants/naaqs-table

in subjects living in urban areas is associated with increased exposure to pollutants, due to several reasons: (a) increased ventilation during exercise increases exposure to pollutants, partly as a result of the shift from nasal to oral ventilation, with loss of nasal filtering function, and (b) the deposition fraction of ultrafine particles increases during exercise [15], especially in children compared to adults and in asthmatic compared to non-asthmatic children [16]. Moreover, exposure during exercise is affected by factors such as time of the day, proximity to road, and traffic intensity. There is evidence of protective behaviors, i.e., decreased physical activity among adults during days or periods of high exposure to air pollutants [17]. The reader is referred to extensive reviews [18, 19] for detailed analysis of the effects of exposure to single pollutants during exercise and in athletes with exercise-induced bronchoconstriction [20]. The following paragraphs will briefly summarize the main effects of pollutants at rest and during exercise.

Ozone: Increased concentration of ozone in breathed air can induce decrements in lung function in healthy young human subjects along with causing airway inflammation [21]. The effect of ozone inhalation on lung function has been extensively evaluated [22–24]. A threshold concentration of 60–80 ppb has been identified for negative health effects of ozone [25]. It is worth noting that the effect of outdoor activities on lung function is controversial: at low levels of exposure, increasing O_3 levels were associated with airway obstruction [26], while lung function tests did not show any consistent pattern of decrease at similar ozone levels even though an increase in exhaled nitric oxide (eNO) concentration was found [27]. Exposure to ozone may produce airway inflammation as evaluated by means of eNO measures [28].

During exercise, effects of ozone exposure may vary in different population subgroups. Höppe and coworkers reported significant decrease in lung function in asthmatic patients and children, whereas elderly subjects or athletes were little affected [29]. A high ozone concentration may impact on athletic performance, but heat and humidity usually associated with ozone pollution appear to play a major role in reducing the performance of endurance-trained runners [30].

Particulate matter: The effects of particulate matter (PM) of different sizes have been extensively studied, with special regard to PM_{10} and $PM_{2.5}$, according to their diameter in μm, as well as ultrafine particles (diameter < 0.1 μm). Healthy subjects exposed to ultrafine carbon particles (without any other component adsorbed) at different concentrations showed a high deposition in the respiratory system, which increased during exercise [31]. During moderate exercise in adult men with prior myocardial infarction, brief exposure to dilute diesel exhaust promoted ST-segment depression, i.e., an important predictor of adverse cardiovascular events [32]. Even mild exercise in a polluted environment may cause increase in biomarkers of airway inflammation and decreased lung function. Adult subjects affected by mild to moderate bronchial asthma walked in Oxford Street, a busy street in London, or in Hyde Park, demonstrating that the degree of traffic exposure may interfere with daily activities in real life [33]. Similarly, in the same ambient conditions, short-term exposure to traffic prevented the beneficial cardiopulmonary effects of walking in subjects affected by chronic obstructive pulmonary disease or ischemic heart disease as well as in healthy individuals [34]. Exposure to diesel exhaust particles (300 $\mu g/m^3$ of $PM_{2.5}$) during moderate or intense exercise in fit young subjects did not acutely modify endothelial function or blood endothelin-1 levels but increased plasma nitrate and nitrite (NOx) concentrations, while exercise intensity did not affect the results [35]. At present, there is suggestive evidence of causal relationships between exposure to traffic-related air pollution and impaired lung function [36].

A current hot topic regards the possible detrimental role of air pollution in healthy people using active transport strategies, i.e., cycling and walking, to commute to study or work as opposed to the use of private cars. The advantages and disadvantages of active transport are being extensively studied, since active transport promotes healthy behavior in the population while decreasing overall traffic volume. Current evidence underlines the pros of active transport [37], although the health benefits of physical exercise appear modified in polluted compared to clean-air areas [38–40].

Figure 14.1 shows an interesting model, where all-cause mortality risk is calculated for air pollution exposure during active transport, with special attention to two points. The "tipping point" represents the amount of cycling per day associated with the highest benefit in terms of risk reduction; the "break-even point "represents the amount of cycling per day beyond which increasing the amount of cycling is associated with increased pollution-related risk [41]. Authors calculated that physical activity should be shortened in very heavily polluted cities, which however represent a low percentage of urban sites in the world [41]. In more complex models, the chronic exposure to pollutants was the main determinant of reduced heart rate variability and increased diastolic blood pressure, whereas the amount of exercise did not exert any significant effect [42]. Therefore, the evidence available to date indicates no additional harm associated with active transport, unless heavy pollution situations occur or cycling is prolonged for many hours per day [41].

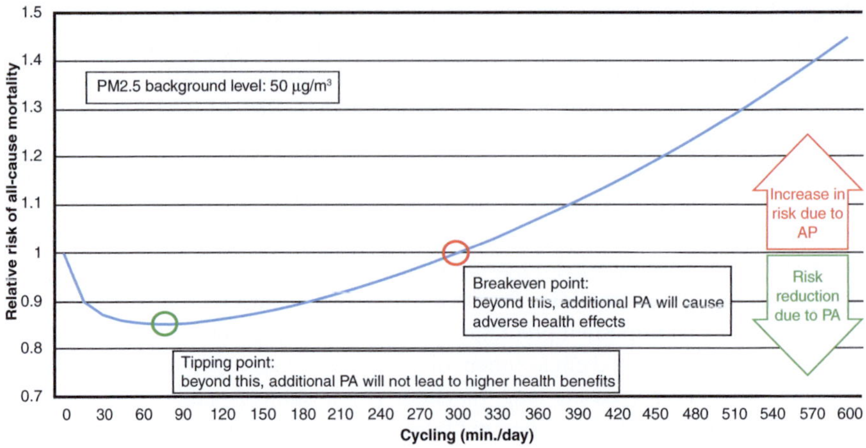

Fig. 14.1 All-cause mortality risk calculated for air pollution exposure during cycling (active transport). The "tipping point" represents the amount of cycling per day associated with the highest benefit in terms of risk reduction; the "break-even point" represents the amount of cycling per day beyond which increasing the amount of cycling is associated with increased pollution-related risk [41]

NO_2: Despite a striking discordance between positive epidemiological associations and results from controlled clinical NO_2 exposure, subjects affected by asthma or allergic diseases are the subgroups most susceptible to NO_2-induced pulmonary effects. Pulmonary adverse effects are not frequent without a co-exposure to specific or nonspecific stimuli [43]. In this context, while studies continue to provide evidence of short-term associations between NO_2 and respiratory outcomes, it has been demonstrated that these associations are not confounded by co-pollutants, including PM and other gaseous pollutants typically used in multipollutant analyses [21]. NO_2 has been used as a marker of air pollution in the Danish study on elderly urban residents, showing positive effects of outdoor physical activity such as cycling and gardening that were especially evident in subjects exposed to low-to-moderate residential NO_2 levels [44].

Sulfur dioxide (SO_2): Inhalation of sulfur dioxide during exercise may produce decrements in lung function [45] with a dose-response relationship in asthmatic volunteers [46].

Carbon monoxide (CO): Studies performed on healthy subjects and patients affected by ischemic heart disease demonstrated that exposure to low-level mixture of CO is able to reduce the exercise time in normal subjects [47] and in patients with coronary artery disease and stable exertional angina [48]. Exercise performance decreased also in COPD patients exposed to 100 ppm of CO for 1 h [49]. In normal young subjects, performing constant power exercise at 85% of maximal O_2 consumption, exposure to 18.9 ppm of CO for 2 h did not affect cardiorespiratory variables but decreased muscle oxygenation assessed by near-infrared spectroscopy (NIRS) [50]. Such an effect was partly reversed by O_2 administration for 1 h after CO exposure [50].

14.3 Athletes and Air Pollution

14.3.1 Data from the Olympic Games

The Olympic Games are major events with large participation of athletes and public. Attention to environmental issues has been far from ideal, with the potential of detrimental effects not only on health but also on athletic performance [51]. For example, during Olympic Games in Beijing 2008, one of the most polluted cities in the world, despite major efforts by authorities to reduce air pollution, i.e., traffic limitation and shutdown of industrial plants, air quality was suboptimal [52]. Similarly, traffic-related air pollution and poor water quality in Guanabara Bay were major problems in Rio de Janeiro 2016 [53]. Monitoring of air quality during Turin Winter Games 2006 showed increased benzene concentrations both in the city and at 1500 m altitude, associated with increased vehicle traffic during the Olympic period [54]. The next Summer Olympic Games in Japan 2020 promise a better air quality and will likely increase the well-being of participating athletes [55].

14.3.2 Which Sports Carry a High Air Pollution Risk?

Analysis of pro-inflammatory effects of air pollutants in athletes is complicated by the effects of exercise per se on airway inflammation, especially in athletes of endurance sports [20, 56]. We have studied the effects of environmental conditions and pollutant concentrations in amateur runners before and after running races at different times of the year. Although during the races pollutant concentrations were below the alert thresholds, ozone concentration was highest in summer. Results showed that apoptosis of airway neutrophils in runners was directly proportional to pollutant concentrations, whereas apoptosis of bronchial epithelial cells appeared mostly affected by intense exercise [57].

There are surprisingly few studies on markers of airway inflammation in athletes exposed to pollutants. In non-elite athletes running for 20 min along a roadway, increased blood levels of toluene, ethylbenzene, and xylenes were documented postexercise [58]. Cavalcante de Sá and coworkers reported that exhaled breath condensate (EBC) pH increased after 5 days in amateur runners training in a clean-air environment (forest), but not in those exposed to air pollution (street). Moreover, impaired nasal mucociliary clearance occurred more often after street than forest training [59]. No differences were found in serum interleukin (IL)-6 or interleukin-10 concentrations, highlighting the lack of sensitive and reliable markers of inflammatory damage in athletes. No changes in breath pH after exercise under clean-air and high ozone and high-PM conditions had been reported by a previous study in high school athletes, but pH in this group was lower than in adult sedentary controls [60].

Some athletes are constantly exposed to pollutants/irritants during training and may develop respiratory symptoms. The best examples are swimmers, who are exposed to chlorine derivatives in indoor swimming pools, and athletes active in ice arenas (hockey players, skaters) who are exposed to CO or NO_2 secondary to malfunction in ice resurfacing process and ultrafine particles [61–63]. The reader is referred to comprehensive reviews dedicated to these topics [20, 64]. Some authors have suggested that high training volumes in elite athletes might justify classification of airway dysfunction as an occupational disease [65]. Technology is under development to decrease high chlorine exposures in swimming pools [66].

14.3.3 Prevention of Negative Effects of Air Pollution

Several simple, common-sense strategies can be adopted to limit the impact of air pollution on health, especially during exercise [67]. These include:

1. Avoiding exercising along high-traffic roads or taking alternative bicycle paths far from traffic [68].
2. Avoiding exercise during the middle of the day, when ozone concentration is highest, especially during summer.
3. Getting real-time information on pollutant levels, wherever available.
4. Training indoor under condition of severe air pollution.
5. Using of protective masks during exercise.

The problem of prevention is particularly relevant in China, where air quality is still far from ideal, children and elderly are used to exercise outdoor, there are no public plans for prevention to pollutant exposures, and the indoor facilities are insufficient to cover the needs of a huge population [69].

There is some evidence that a healthy diet and vitamin C and D supplementation can help to limit air pollution-related respiratory damage [70]. Interestingly, a recent experimental study in mice reported a role of gut microbiome in the pathogenesis of increased bronchial reactivity in response to ozone exposure [71]. This is a new and poorly explored field, which deserves further study.

14.4 Conclusions

Figure 14.2 tries to summarize the current knowledge on the effects of exercise under conditions of air pollution in athletes. While there is evidence that exercise does potentiate exposure to pollutants, the possibility that healthy well-trained athletes may develop protective defense mechanisms has been not explored to date. Exercise training is associated with pro- and anti-inflammatory changes, and a heightened response to oxidative stress as a result of training may well contribute to limit the damage linked to air pollution. Much work needs to be done in this respect, as well as studies on air pollutant thresholds which may affect performance.

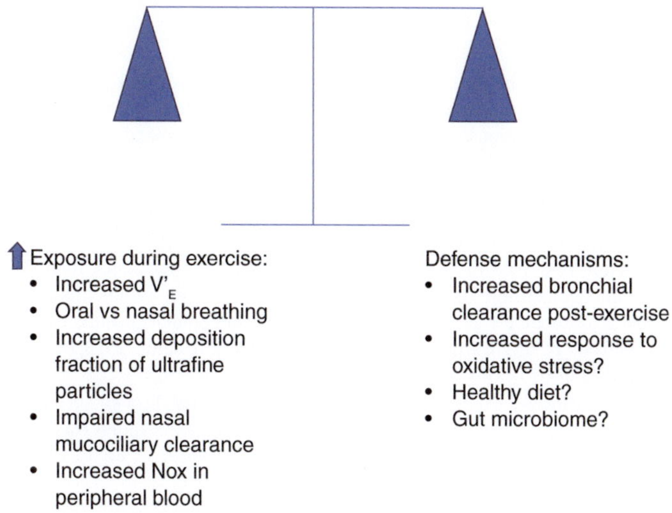

Fig. 14.2 Summary of mechanisms of increased exposure to pollutant during exercise in athletes. The impact of potential defense mechanisms has not been studied in detail

Air quality has improved in the last decades worldwide due to increased knowledge about health risks and implementation of preventive measures. Current evidence shows that the benefits of exercise in healthy subjects outweigh the effects of concurrent air pollution, suggesting that preventive strategies against exposure to pollutants should not become a recommendation for a sedentary lifestyle. Vulnerable populations, such as children, elderly, and patients with cardiorespiratory diseases, should refrain from exercising under air pollution conditions, as indicated by studies showing increasing mortality and morbidity risk.

Key Points

- Air pollution is a threat for human health, and exercise increases exposure to pollutants
- The benefits of physical exercise can be decreased by air pollution, but exercise-associated risk reduction in healthy subjects remains significant unless severe air pollution occurs
- Vulnerable groups such as children and patients with cardiorespiratory disease should avoid exercise in polluted areas
- Little is known on the effects of air pollution in athletes and on possible protective mechanisms evoked by habitual exercise which may help to limit the detrimental effects of air pollution on cardiorespiratory health

References

1. United Nations. World urbanization prospects: the 2014 revision, highlights. New York: United Nations, Department of Economic and Social Affairs, Population Division; 2014.. https://esa.un.org/unpd/wup/Publications/Files/WUP2014-Report.pdf. Accessed 13 May 2018
2. Mueller N, Rojas-Rueda D, Basagaña X, Cirach M, Cole-Hunter T, Dadvand P, Donaire-Gonzalez D, Foraster M, Gascon M, Martinez D, Tonne C, Triguero-Mas M, Valentín A, Nieuwenhuijsen M. Urban and transport planning related exposures and mortality: a health impact assessment for cities. Environ Health Perspect. 2017;125:89–96.
3. Brook JR, Setton EN, Seed E, Shooshtari M, Doiron D, The Canadian Urban Environmental Health Research Consortium (CANUE). The Canadian urban environmental Health Research consortium- a protocol for building a national environmental exposure data platform for integrated analysis of urban from and health. BMC Public Health. 2018;18:114. https://doi.org/10.1186/s12889-017-5001-5.
4. Götschi T, de Nazelle A, Brand C, Gerike R, Physical activity through sustainable transport approaches (PASTA) consortium. Towards a comprehensive conceptual framework of active travel behavior: a review and synthesis of published frameworks. Curr Environ Health Rep. 2017;4:286–95.
5. Vedal S, Kaufman JD. What does multi-pollutant air pollution research mean? Am J Respir Crit Care Med. 2011;183:4–6.
6. Jerrett M, Donaire-Gonzalez D, Popoola O, Jones R, Cohen RC, Almanza E, de Nazelle A, Mead I, Carrasco-Turigas G, Cole-Hunter T, Triguero-Mas M, Seto E, Nieuwenhuijsen M. Validating novel air pollution sensors to improve exposure estimates for epidemiological analyses and citizen science. Environ Res. 2017;158:286–94.
7. Dons E, Laeremans M, Orjuela JP, Avila-Palencia I, Carrasco-Turigas G, Cole-Hunter T, Anaya-Boig E, Standaert A, De Boever P, Nawrot T, Götschi T, de Nazelle A, Nieuwenhuijsen M, Int Panis L. Wearable sensors for personal monitoring and estimation of inhaled traffic-related air pollution: evaluation of methods. Environ Sci Technol. 2017;51:1859–67.
8. Requia WJ, Adams MD, Arain A, Papatheodorou S, Koutrakis P, Mahmoud M. Global association of air pollution and cardiorespiratory diseases: a systematic review, meta-analysis, and investigation of modifier variables. Am J Public Health. 2018;108(S2):S123–30.
9. Heinrich J, Schikowski T. COPD patients as vulnerable subpopulation for exposure to ambient air pollution. Curr Environ Health Rep. 2018;5:70–6.
10. Airnow. Air quality index- a guide to air quality and your health. https://airnow.gov/index.cfm?action=aqi_brochure.index. Accessed 13 May 2018.
11. WHO Expert Consultation: Available evidence for the future update of the WHO Global Air Quality Guidelines (AQGs). http://www.euro.who.int/__data/assets/pdf_file/0013/301720/Evidence-future-update-AQGs-mtg-report-Bonn-sept-oct-15.pdf. Accessed 13 May 2018.
12. McDonnell WF, Stewart PW, Smith MV. Prediction of ozone-induced lung function responses in humans. Inhal Toxicol. 2010;22:160–8.
13. Viegi G, Baldacci S. Epidemiological studies of chronic respiratory conditions in relation to urban air pollution in adults. Eur Respir Mon. 2002;21:1–16.
14. Cromar KR, Gladson LA, Ghazipura M, Ewart G. Estimated excess morbidity and mortality associated with air pollution above American Thoracic Society-recommended Standards, 2013–2015. American Thoracic Society and Marron Institute Report. Ann Am Thorac Soc. 2018;15:542–51.
15. Blair SN. Physical inactivity: the biggest public health problem of the 21st century. Br J Sports Med. 2009;43:1–2.
16. Daigle CC, Chalupa DC, Gibb FR, Morrow PE, Oberdörster G, Utell MJ, Frampton MW. Ultrafine particle deposition in humans during rest and exercise. Inhal Toxicol. 2003;15:539–52.
17. Olvera HA, Perez D, Clague JW, Cheng Y-S, Li W-W, Amaya MA, Burchiel SW, Berwick M, Pingitore NE. The effect of ventilation, age and asthmatic condition on ultrafine particle deposition in children. Pulm Med. 2012;2012:736290. https://doi.org/10.1155/2012/736290.

18. An R, Zhang S, Ji M, Guan C. Impact of ambient air pollution on physical activity among adults: a systematic review and meta-analysis. Persp Public Health. 2018;138:111–21.
19. Carlisle AJ, Sharp NCC. Exercise and outdoor ambient air pollution. Br J Sports Med. 2001;35:214–22.
20. Giles LV, Koehle MS. The health effects of exercising in air pollution. Sports Med. 2014;44:223–49.
21. Rundell KW, Anderson SD, Sue-Chu M, Bougault V, Boulet LP. Air quality and temperature effects on exercise-induced bronchoconstriction. Compr Physiol. 2015;5:579–610.
22. WHO Regional Office for Europe. Review of evidence on health aspects of air pollution—REVIHAAP project: technical report. Copenhagen, WHO Regional Office for Europe. 2013. http://www.euro.who.int/__data/assets/pdf_file/0004/193108/REVIHAAP-Final-technical-report.pdf. Accessed 13 May 2018.
23. McDonnell WF, Stewart PW, Smith MV. The temporal dynamics of ozone-induced FEV1 changes in humans: an exposure-response model. Inhal Toxicol. 2007;19:483–94.
24. McDonnell WF, Stewart PW, Smith MV, Kim CS, Schelegle ES. Prediction of lung function response for populations exposed to a wide range of ozone conditions. Inhal Toxicol. 2012;24:619–33.
25. Schelegle ES, Morales CA, Walby WF, Marion S, Allen RP. 6.6-hour inhalation of ozone concentrations from 60 to 87 parts per billion in healthy humans. Am J Respir Crit Care Med. 2009;180:265–72.
26. Thaller EI, Petronella SA, Hochman D, Howard S, Chhikara RS, Brooks EG. Moderate increases in ambient PM2.5 and ozone are associated with lung function decreases in beach lifeguards. J Occup Environ Med. 2008;50:202–11.
27. Nickmilder M, de Burbure C, Carbonnelle S, Dumont X, Bernard A, Derouane A. Increase of exhaled nitric oxide in children exposed to low levels of ambient ozone. J Toxicol Environ Health A. 2007;70:270–4.
28. Berhane K, Zhang Y, Linn WS, Rappaport EB, Bastain TM, Salam MT, Islam T, Lurmann F, Gilliland FD. The effect of ambient air pollution on exhaled nitric oxide in the Children's Health Study. Eur Respir J. 2011;37:1029–36.
29. Höppe P, Peters A, Rabe G, Praml G, Lindner J, Jakobi G, Fruhmann G, Nowak D. Environmental ozone effects in different population subgroups. Int J Hyg Environ Health. 2003;206:505–16.
30. Gomes EC, Stone V, Florida-James G. Investigating performance and lung function in a hot, humid, and ozone-polluted environment. Eur J Appl Physiol. 2010;110:199–205.
31. Frampton MW. Systemic and cardiovascular effects of airway injury and inflammation: ultrafine particle exposure in humans. Environ Health Perspect. 2001;109(Suppl. 4):529–32.
32. Mills NL, Törnqvist H, Gonzalez MC, Vink E, Robinson SD, Söderberg S, Boon NA, Donaldson K, Sandström T, Blomberg A, Newby DE. Ischemic and thrombotic effects of dilute diesel-exhaust inhalation in men with coronary heart disease. N Engl J Med. 2007;357:1075–82.
33. McCreanor J, Cullinan P, Nieuwenhuijsen MJ, Stewart-Evans J, Malliarou E, Jarup L, Harrington R, Svartengren M, Han IK, Ohman-Strickland P, Chung KF, Zhang J. Respiratory effects of exposure to diesel traffic in persons with asthma. N Engl J Med. 2007;357: 2348–58.
34. Sinharay R, Gong J, Barratt B, Ohman-Strickland P, Ernst S, Kelly FJ, Zhang JJ, Collins P, Cullinan P, Chung KF. Respiratory and cardiovascular responses to walking down a traffic-polluted road compared with walking in a traffic-free area in participants aged 60 years and older with chronic lung or heart disease and age-matched healthy controls: a randomised, crossover study. Lancet. 2018;391:339–49.
35. Giles LV, Tebbutt SJ, Carlsten C, Koehle MS. The effect of low and high-intensity cycling in diesel exhaust on flow-mediated dilation, circulating NOx, endothelin-1 and blood pressure. PLoS One. 2018;13(2):e0192419.
36. Health Effects Institute. Panel on the Health Effects of Traffic-Related Air Pollution. Traffic-related air pollution: a critical review of the literature on emissions, exposure, and health effects. 2010. https://www.healtheffects.org/publication/traffic-related-air-pollution-critical-review-literature-emissions-exposure-and-health. Accessed 13 May 2018.

37. Mueller N, Rojas-Rueda D, Cole-Hunter T, de Nazelle A, Dons E, Gerike R, Götschi T, Int Panis L, Kahlmeier S, Nieuwenhuijsen MJ. Health impact assessment of active transportation: a systematic review. Prev Med. 2015;76:103–14.
38. Matt F, Cole-Hunter T, Donaire-Gonzales D, Kubesch N, Martinez D, Carrasco-Turigas G, Nieuwenhuijsen M. Acute respiratory response to traffic-related air pollution during physical activity performance. Environ Int. 2016;97:45–55.
39. Laeremans M, Dons E, Avila-Palencia I, Carrasco-Turigas G, Orjuela JP, Anaya E, Cole-Hunter T, de Nazelle A, Nieuwenhuijsen M, Standaert A, Van Poppel M, De Boever P, Int Panis L. Short-term effects of physical activity, air pollution and their interaction on the cardiovascular and respiratory system. Environ Int. 2018;117:82–90.
40. Laeremans M, Dons E, Avila-Palencia I, Carrasco-Turigas G, Orjuela-Mendoza JP, Anaya-Boig E, Cole-Hunter T, de Nazelle A, Nieuwenhuijsen M, Standaert A, Van Poppel M, De Boever P, Int Panis L. Black carbon reduces the beneficial effect of physical activity on lung function. Med Sci Sports Exerc. 2018;50(9):1875–81. https://doi.org/10.1249/MSS.0000000000001632.
41. Tainio M, de Nazelle AJ, Götschi T, Kahlmeier S, Rojas-Rueda D, Nieuwenhuijsen MJ, Hérick de Sa T, Kelly P, Woodcock J. Can air pollution negate the health benefits of cycling and walking? Prev Med. 2016;87:233–6.
42. Cole-Hunter T, de Nazelle A, Donaire-Gonzales D, Kubesch N, Carrasco-Turigas G, Matt F, Foraster M, Martinez D, Ambros A, Cirach M, Martinez D, Belmonte J, Nieuwenhuijsen M. Estimated effects of air pollution and space-time-activity on cardiopulmonary outcomes in healthy adults: a repeated measures study. Environ Int. 2018;111:247–59.
43. Hesterberg TW, Bunn WB, McClellan RO, Hamade AK, Long CM, Valberg PA. Critical review of the human data on short-term nitrogen dioxide (NO_2) exposures: evidence for NO_2 no-effect levels. Crit Rev Toxicol. 2009;39:743–81.
44. Andersen ZJ, de Nazelle A, Mendez MA, Garcia-Aymerich J, Hertel O, Tjønneland A, Overvad K, Raaschou-Nielsen O, Nieuwenhuijsen MJ. A study of the combined effects of physical activity and air pollution on mortality in elderly urban residents: the Danish Diet, Cancer, and Health Cohort. Environ Health Perspect. 2015;123:557–63.
45. EPA integrated science assessment for oxides of nitrogen – health criteria. https://ofmpub.epa.gov/eims/eimscomm.getfile?p_download_id=526855. Accessed 13 May 2018.
46. Johns DO, Svendsgaard D, Linn WS. Analysis of the concentration-respiratory response among asthmatics following controlled short-term exposures to sulfur dioxide. Inhal Toxicol. 2010;22:1184–93.
47. Koike A, Wasserman K. Effect of acute reduction in oxygen transport on parameters of aerobic function during exercise. Ann Acad Med Singap. 1992;21:14–22.
48. Allred EN, Bleecker ER, Chaitman BR, Dahms TE, Gottlieb SO, Hackney JD, Pagano M, Selvester RH, Walden SM, Warren J. Effects of carbon monoxide on myocardial ischemia. Environ Health Perspect. 1991;91:89–132.
49. Aronow WS, Ferlinz J, Glauser F. Effect of carbon monoxide on exercise performance in chronic obstructive pulmonary disease. Am J Med. 1977;63:904–8.
50. Keramidas ME, Kounalakis SN, Eiken O, Mekjavic IB. Carbon monoxide exposure during exercise performance: muscle and cerebral oxygenation. Acta Physiol. 2012;204:544–54.
51. Fitch K. Air pollution, athletic health and performance at the Olympic Games. J Sports Med Phys Fitness. 2016;56:922–32.
52. Schleicher N, Norra S, Chen Y, Chai F, Wang S. Efficiency of mitigation measures to reduce particulate air pollution-A case study during the Olympic Summer Games 2008 in Beijing, China. Sci Total Environ. 2012;42:146–58.
53. Gioda A, Ventura LMB, Ramos MB, Silva MPR. Half century monitoring air pollution in a megacity: a case study of Rio de Janeiro. Water Air Soil Pollut. 2016;227:1–17.
54. Bono R, Degan R, Pazzi M, Romanazzi V, Rovere R. Benzene and formaldehyde in air of two Winter Olympic venues of "Torino 2006". Environ Int. 2010;36:269–75.
55. Donnelly AA, MacIntyre TE, O'Sullivan N, Warrington G, Harrison AJ, Igou ER, Jones M, Gidlow C, Brick N, Lahart I, Cloak R, Lane AM. Environmental influences on elite sport

athletes well being: from gold, silver, and bronze to blue green and gold. Front Psychol. 2016;7:1167. https://doi.org/10.3389/fpsyg.2016.01167.
56. Bonsignore MR, Morici G, Vignola AM, Riccobono L, Bonanno A, Profita M, Abate P, Scichilone N, Amato G, Bellia V, Bonsignore G. Increased airway inflammatory cells in endurance athletes: what do they mean? Clin Exp Allergy. 2003;33:14–21.
57. Chimenti L, Morici G, Paterno A, Bonanno A, Vultaggio M, Bellia V, Bonsignore MR. Environmental conditions, air pollutants, and airway cells in runners: a longitudinal field study. J Sports Sci. 2009;27:925–35.
58. Blair C, Walls J, Davies NW, Jacobson GA. Volatile organic compounds in runners near a roadway: increased blood levels after short-duration exercise. Br J Sports Med. 2010;44:731–5.
59. Cavalcante de Sá M, Nakagawa NK, Saldiva de André CD, Carvalho-Oliveira R, de Santana Carvalho T, Nicola ML, de André PA, Nascimento Saldiva PH, Vaisberg M. Aerobic exercise in polluted urban environments: effects on airway defense mechanisms in young healthy amateur runners. J Breath Res. 2016;10:046018.
60. Ferdinands JM, Crawford CA, Greenwald R, Van Sickle D, Hunter E, Teague WG. Breath acidification in adolescent runners exposed to atmospheric pollution: a prospective, repeated measures observational study. Environ Health. 2008;7:10. https://doi.org/10.1186/1476-069X-7-10.
61. Creswell PD, Meiman JG, Nehls-Lowe H, Vogt C, Wozniak RJ, Werner MA, Anderson H. Exposure to elevated carbon monoxide levels at an indoor ice arena--Wisconsin, 2014. MMWR Morb Mortal Wkly Rep. 2015;64:1267–70.
62. Brat K, Merta Z, Plutinsky M, Skrickova J, Stanek M. Ice hockey lung – a case of mass nitrogen dioxide poisoning in the Czech Republic. Can Respir J. 2013;20(6):e100–3.
63. Rundell KW. Pulmonary function decay in women ice hockey players: is there a relationship to ice rink air quality? Inhal Toxicol. 2004;16:117–23.
64. Mountjoy M, Fitch K, Boulet LP, Bougault V, van Mechelen W, Verhagen E. Prevalence and characteristics of asthma in the aquatic disciplines. J Allergy Clin Immunol. 2015;136:588–94.
65. Price OJ, Ansley L, Menzies-Gow A, Cullinan P, Hull JH. Airway dysfunction in elite athletes – an occupational lung disease? Allergy. 2013;68:1343–52.
66. Gomà A, de Lluis R, Roca-Ferrer J, Lafuente J, Picado C. Respiratory, ocular and skin health in recreational and competitive swimmers: beneficial effect of a new method to reduce chlorine oxidant derivatives. Environ Res. 2017;152:315–21.
67. Laumbach R, Meng Q, Kipen H. What can individuals do to reduce personal health risks from air pollution? J Thorac Dis. 2015;7:96–107.
68. Cole-Hunter T, Jayaratne R, Stewart I, Hadaway M, Morawska L, Solomon C. Utility of an alternative bicycle commute route of lower proximity to motorized traffic in decreasing exposure to ultra-fine particles, respiratory symptoms and airway inflammation--a structured exposure experiment. Environ Health. 2013;12(1):29.
69. Li F, Liu Y, Lü J, Linag L, Harmer P. Ambient air pollution in China poses a multifaceted health threat to outdoor physical activity. J Epidemiol Commun Health. 2015;69:201–4.
70. Whyand T, Hurst JR, Beckles M, Caplin ME. Pollution and respiratory disease: can diet or supplements help? A review. Respir Res. 2018;19(1):79.
71. Cho Y, Abu-Ali G, Tashiro H, Kasahara DI, Brown TA, Brand JD, Mathews JA, Huttenhower C, Shore SA. The microbiome regulates pulmonary responses to ozone in mice. Am J Respir Cell Mol Biol. 2018;59(3):346–54. https://doi.org/10.1165/rcmb.2017-0404OC.

MIX
Papier aus verantwortungsvollen Quellen
Paper from responsible sources
FSC® C105338
FSC
www.fsc.org

If you have any concerns about our products,
you can contact us on
ProductSafety@springernature.com

In case Publisher is established outside the EU,
the EU authorized representative is:
Springer Nature Customer Service Center GmbH
Europaplatz 3, 69115 Heidelberg, Germany

Printed by Libri Plureos GmbH
in Hamburg, Germany